Wilhelm Schäfer (Ed.)

Software Process Technology

4th European Workshop, EWSPT '95
Noordwijkerhout, The Netherlands, April 3-5, 1995
Proceedings

Springer

Series Editors

Gerhard Goos
Universität Karlsruhe
Vincenz-Priessnitz-Straße 3, D-76128 Karlsruhe, Germany

Juris Hartmanis
Department of Computer Science, Cornell University
4130 Upson Hall, Ithaca, NY 14853, USA

Jan van Leeuwen
Department of Computer Science, Utrecht University
Padualaan 14, 3584 CH Utrecht, The Netherlands

Volume Editor

Wilhelm Schäfer
FB 17, Universität-GH-Paderborn
D-33095 Paderborn

CR Subject Classification (1991): D.2, K.6, K.4.2

ISBN 3-540-59205-9 Springer-Verlag Berlin Heidelberg New York

CIP data applied for

© Springer-Verlag Berlin Heidelberg 1995
Printed in Germany

Typesetting: Camera-ready by author
SPIN: 10485723   06/3142-543210 - Printed on acid-free paper

Preface

Software Process Technology (SPT) has increasingly become a key technology to cope with the challenges of large team-oriented software production. In addition, as software intensive systems have become an integral part of our daily life, very high quality demands are put on the software as well as on its production processes.

SPT especially develops the technical foundations to make the software production process more reliable, faster and easier to measure and control. It also addresses closely related areas like business process (re-)engineering, workflow management and the like by applying results from SPT in those areas and vice versa.

It is in this context that the European Union sponsored a Basic Research Action Working Group called PROMOTER (Process Modeling Techniques Basic Research). This group has been in place since September 1992 and one of its results was to establish a Europe-based workshop series on SPT called the European Workshop on Software Process Technology (EWSPT). The current edition is already the fourth in this series with the fifth coming up in 1996. It will be held in Nancy (France).

The international establishment of the series is indicated by the large number of requests we got from people who wished to participate when they heared about the workshop. We also have a fair number of non-European attendees whose papers were accepted.

Overall, we got again about 50 submissions from around the world. The Programme Committee accepted about half of the submissions to keep the number of attendees small enough to enable intensive, focussed discussions.

For all the preparatory work done so far including paper reviewing, session preparation, etc., I am particulary thankful to the members of the Programme Committee: Reidar Conradi, Jean-Claude Derniame, Christer Fernstrøm, Alfonso Fuggetta, Volker Gruhn, Nazim Madhavji, Carlo Montangero, Henk Obbink, Colin Tully, Brian Warboys, and Gregor Engels who as a general chair together with his team also did a splendid job on the local organisation. Finally, we acknowledge the professional support from Springer, who are publishing these proceedings again. Proceedings of the two previous workshops were also published as Springer LNCS.

Based on the submitted and accepted papers the Programme Committee designed sessions under the following topics: analysis and metrics, application
experiments, language experiments, related domains, models for distributions,
mechanisms for cooperation, and change and meta-process. Those session
topics also form the chapters of these proceedings. Following the themes
through the workshop series shows the emergence of the field but also identifies the areas which are getting most attention from researchers. I am looking
forward to an active and fruitful workshop.

February 1995   Wilhelm Schäfer
                                                  Programme Committee Chair

Table of Contents

Mechanisms for Cooperation
C. Fernström

Session on Change and Meta-process
R. Conradi...240

Metrics and Analysis Session

J. Henk Obbink

Philips Research Laboratories
Prof. Holstlaan 4, 5656 AA Eindhoven, The Netherlands
e-mail: obbink@prl.philips.nl

This is the first time that a full session in the workshop is devoted to what I would call the deeper issues of process technology. Much of the work on process technology is motivated by referring to the benefits it provides to the improvement of the overall software process. In order to improve on the right issues and in the right direction, we need a deeper analysis than is possible by just having flat- single-view process models. The complexity of real processes just force us to consider multiple viewpoint process models. In order to know quantitatively what the actual progress is we have made, we need means to compare the current situation with the past situation in a quantitative way, which leads us automatically in the metrics area.

In the "Process Viewpoints" paper the need is discussed for a systematic framework which can be used to analyse software processes and derive process models. The notion of process viewpoints is introduced which incorporate questions about process and potential process improvements. The questions are derived from organisational concerns which must be explicitly identified. The Process viewpoint concept is inspired by requirements engineering research and has been actually originated from improvement activities in requirements engineering.

The second paper on "Process-Based Software Risk Assessment" the need for analysing process models to predict the behaviour of software processes is stressed, in order to help in planning and enacting software projects. The reasoning is applied to one of the more difficult or fuzzy processes that of assessing process related risks.

The third paper on "The Use of Roles aand Measurement to Enact Project Plans in MVP-S" stresses the need to incorporate explicitly that these ingredients are an essential prerequisite for systematics process improvement. The need for measurement is motivated and the integration of the measurement and modelling approaches is sketched. This integrated approach is supported by a prototype system which offers process guidance using the role definitions and the measurement data.

The fourth paper on "Combining Process Models and Metrics in Practice" also argues for the need to combine process modelling and metrication in order to provide a sound basis for continuous process improvement. Practical experiences are presented to support this proposal. These experiences have been gained in a research environment and in an industry group working on consumer electronics. Practical experiences carried out so far, show that it is plausible to combine metrication and process modelling and thereby profit from their mutual influence.

Process Viewpoints

Ian Sommerville, Gerald Kotonya, Steve Viller and Pete Sawyer

Computing Dept., Lancaster University, LANCASTER LA1 4YR, UK.
E-mail: is@comp.lancs.ac.uk

Abstract. This position paper discusses the need for a systematic framework which can be used to analyse software processes and derive process models. We propose the idea of process viewpoints which have associated process models and which incorporate questions about process and potential process improvement. The questions associated with each process viewpoint are derived from organisational concerns which must be explicitly identified. This work has been carried out in the context of a project which is investigating approaches to requirements engineering process improvement.

1 Introduction

Over the last ten years, the software process community has focused its attention on the development of process models, process modelling formalisms and methods of enacting these process models to support the development process. Less attention has been paid to the problem of discovering the actual process models which are used. It has been generally assumed that it is relatively straightforward to understand existing processes (although it is recognised that these processes may be complex) and that the principal problems lie in producing realistic models of these processes.

We note a parallel here with the requirements engineering community who based their work for many years on the idea that system requirements were simply floating around to be 'captured' and used as the basis of a system model. We now realise that understanding and discovering system requirements is a very difficult process which is far removed from the idea of fishing in a sea of readily available requirements. We suggest that if informal and unstructured approaches are used to analyse software processes and derive process models, these are unlikely to be successful. Rather, we need to derive systematic approaches to discovering the actual processes used in an organisation and their relationships. The Elicit approach [1] is an example of one of the few systematic approaches which have been developed for process understanding.

The need for a systematic approach to process understanding was confirmed by a number of empirical studies of software processes which we carried out in a number of organisations [2] [3]. We discovered that different participants in the process were working to different process models. We believe that this finding goes some way towards explaining the 'non-conformance' to process models which is very obvious in many organisations. It is not that a published or standardised process model is necessarily incorrect. It is simply that the process model represents only one particular model (usually that of management). Non-conformance to this model reflects the fact that other process participants are working to some different (but not necessarily less 'correct') model.

We do not believe that this problem of non-conformance can be solved by finding a single, all-encompassing model which is acceptable to all who have a direct or indirect interest in the process. Rather, we should accept that multiple process models exist and we should focus on discovering these models and, where appropriate, reconciling them. This is particularly important if a process is to be supported by some process technology or if process improvements are proposed. If improvement proposals do not apply across the different models held by process stakeholders, they are unlikely to be successful.

Ethnographic studies can provided useful insights into software processes but we discovered that they were not the best approach to deriving models of parts of the process. The duration of software processes is too long for ethnographic analysis to cost-effective. We concluded that ethnographic studies had a role to play in process understanding but only if they were used in conjunction with some more systematic framework. This framework should capture the diverse process models held by different process stakeholders and act as a starting point for reconciling these models and improving software processes.

We propose that we should explicitly identify process viewpoints and associate one or more process models with each of these viewpoints. The notion of process viewpoints which we are developing has been influenced by Basili and Rombach's GQM (Goals-Questions-Metrics) approach to process improvement [4] and by work on viewpoints for software requirements elicitation and analysis [5-7].

A process viewpoint is an encapsulated process description which includes a process model, the sources of model information, and the questions which were posed to derive that process model. These questions may be derived from organisational considerations which are common to all viewpoints. These organisational considerations may be specific process activities such as 'the design activity' or may be related to process attributes such as cost or product quality. We call these organisational considerations 'concerns'. The notion of a concern is a critical one and we discuss it in more detail later in the paper.

Concerns are used to stimulate the generation of process questions which may be specific to a particular process viewpoint or which may be posed to all viewpoints. Process questions fall into two principal classes, with some overlap between them:

1. *Exploratory questions* These are intended to discover information about the process which is being studied. The answers to these questions influence the process models associated with a viewpoint. An example of an exploratory question associated with a cost concern (say) would be 'what mechanisms are incorporated in the process to monitor the costs of project activities'.

2. *Improvement questions* These are intended to discover what is required to effect improvement in the process. The answers to these questions should help identify process revisions. An example of an improvement question might be 'how can we reduce the time required to review documents'. Answering improvement questions may result in the creation of new process models.

Process questions are used to drive the process analysis. Answers to these questions may be discovered either by explicitly asking process stakeholders or by observation, studies of process documentation and other material. The process models or associated information (such as process rationale) which are derived should reflect the answers to the questions associated with the process viewpoint.

We are currently applying this approach to studies of the requirements engineering process in two different organisations in a European project called

REAIMS. The overall objective of REAIMS is to provide a framework for requirements engineering process improvement. We are therefore concerned with deriving both exploratory and improvement questions about these processes.

2   A definition of process viewpoints

It has been recognised since the mid-1970s that top-down system analysis is simplistic and that the requirements for a system derive from many different sources. Each of these sources considers the system in different ways (e.g. the driver of a train looks at a signalling system in a different way from the train operating company). It is often the case that the system requirements derived from different sources are inconsistent and conflict in some way.

We argue that a comparable approach should be taken to understanding processes and that such a multi-perspective approach is likely to lead to a deeper understanding of the real process and the needs of the participants in that process. We believe that there is no such thing as a single software process model which will be accepted by all of the stakeholders in the process. Rather, we argue that there are different ways of looking at the process (viewpoints) with different associated process models.

We do not think it useful to consider these different process models to be views of some all-encompassing process model. While it may be possible to integrate all the separate models, the resulting overall model is likely to be so complex that it will be completely incomprehensible. It may never, therefore, be possible to produce a 'complete' model of the process. We do not see this as a problem so long as we can define and manage interfaces between the different models held by different viewpoints.

We consider a process viewpoint to be an encapsulation of process information. It may be modelled as a septuple as follows:

Viewpoint = { Name, Scope, Models, Concerns, Organisational questions,
Local questions, Sources }

Name
   The name of a viewpoint is a meaningful term used to refer to the viewpoint.

Scope
   The scope of a viewpoint is a specification of the limits of that viewpoint i.e. it defines the focus of a viewpoint on a particular process. For example, the scope of a viewpoint may be the accounting function. It would therefore be expected that associated models would focus on resource utilisation. An explicit identification of scope helps us to understand why models have been formulated in a particular way.

Models
   A viewpoint may have one or more associated process models. These can be in any appropriate notation from natural language descriptions to formal mathematical text. This flexibility is essential as there is no single formal notation which could be understood by all viewpoints.

Concerns
   Each viewpoint has an associated set of concerns which are used to drive the process of process understanding, modelling and improvement. Typical concerns might be cost reduction, improved time to delivery, increased process visibility, etc. Concerns must be addressed by all viewpoints.

Organisational questions

Each viewpoint has an associated set of organisational questions which must be addressed as part of the process modelling and improvement process. Organisational questions are those questions which constrain or influence local questions derived in a viewpoint. For example, an organisational question might be 'what is the relationship between all of the product development processes in the organisation'. Each viewpoint should then address this question and refine it to more specific local questions. Organisational questions are usually derived from local questions generated in other viewpoints.

Local questions

Each viewpoint has an associated set of local questions which may be refinements of the organisational questions or which may be separate questions in their own right. Local questions, therefore, may develop or reword organisational questions so that they are appropriately formulated for that viewpoint.

Sources

Each viewpoint has an associated set of sources (people or documents) which provide the information associated with the viewpoint. The explicit maintenance of sources allows us to trace information and to know who to negotiate with when conflicts and disagreements arise.

We classify viewpoints into two groups namely:

1. *Direct viewpoints* These are associated with participants in the process such as designers, programmers, test engineers, etc. Viewpoints, however, are not normally mapped on a 1-1 basis to roles. Rather, they would normally be associated with teams (e.g. a testing team) which may encompass a number of different roles.

2. *Indirect viewpoints* These are associated with organisations and customers who may influence the process used but who do not actively participate in it.

It is important to emphasise that viewpoints are ways of looking at a process and that the same person can look at a process in quite different ways. For example, project managers can consider a process from a technical (direct) viewpoint if they are interested in the activities undertaken by the project development team. They can also take an organisational (indirect) viewpoint when considering issues of process management.

The classification into direct and indirect viewpoints is useful because it recognises the inherent tension between user-centred models and organisational models. We know that published organisational process models often do not reflect reality. The actual process followed by software engineers is quite different from these published models. The question we seek to answer is why the organisational models are so different. By collecting both organisational and participant models, we hope to discover conflicts and discrepancies and hence understand the relationships between these different models.

3 Concerns

The notion of a concern is an important one and it is worth explaining it in a bit more detail here. Basically, a concern is an organisational issue which must be considered by all viewpoints irrespective of whether they are direct or indirect viewpoints. The term 'concern' is used intuitively in that an organisation may be *concerned* with issues such as the cost of a process, the design methods which are used or the interaction between different teams involved in the process. Of course, a viewpoint may decide that a particular concern is irrelevant but it should make and document this choice explicitly rather than simply ignore that concern.

Figure 1 shows the orthogonality of viewpoints and concerns. The actual enacted process is at the apex of the triangle so that as the viewpoint moves from the apex to the base, it becomes more and more remote from the process itself. However, remoteness is not the same as irrelevance. Indirect viewpoints often have much more political power than direct viewpoints or organisational viewpoints which are close to the process. Therefore, a large customer may mandate a process model which must be used (or which, at least, the organisation must appear to use) irrespective of the organisational models which are already in place.

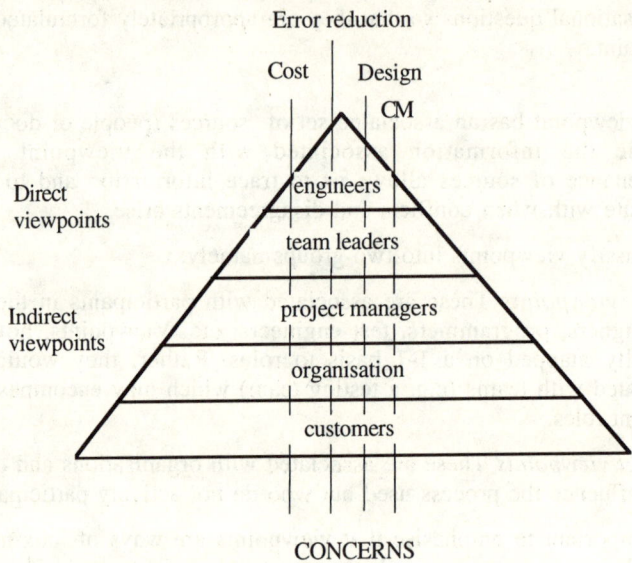

Figure 1 Viewpoints and concerns

There are two principal types of concern:

1. *Sub-processes* A concern is a particular sub-process. Examples of these concerns would be 'testing', 'design', 'configuration management', etc.

2. *'Non-functional' concerns* These are concerns such as cost, quality, time to delivery etc. They are not generally considered in isolation but in conjunction with other concerns.

For each-sub-process concern, each viewpoint may have an associated process model (Figure 2). Part of the analysis process considers each concern for all viewpoints and compares the process models. Discrepancies in these models suggest potential process problems and misunderstandings between participants and process stakeholders.

Concerns are used to generate questions which may be either local questions or which may be broader questions concerning organisation issues. These broader organisational questions are exported from a viewpoint to all other viewpoints. Therefore, all viewpoints share a common set of organisational questions which are developed during an initial analysis. Because organisational questions do not all emerge at once, this means that the analysis process is necessarily iterative; as new organisational questions emerge, each viewpoint has to re-enter the analysis activity. Clearly, some tool support will be needed for this but we have not yet looked at what level of support can be economically provided.

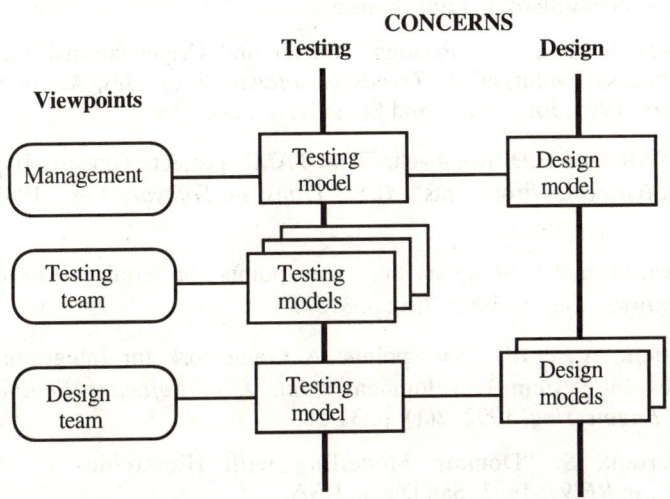

Figure 2 Concerns and models

As a general rule, all concerns should either be explicitly rejected by a viewpoint or should have at least one question associated with them. We have not yet developed a process for using the viewpoints approach but we anticipate that question generation from concerns will be a key part of this process.

4   Conclusions

The proposed approach to process analysis has been prompted by our previous empirical process studies which revealed considerable process complexity and diversity. They also revealed the high costs of these empirical studies and the difficulties of applying the output from these studies in an effective way. We have therefore developed this more structured approach which we hope will reduce the costs of analysis and provide us with more rapidly usable information.

The work described here is still at an early stage of development and we have not yet attempted to apply it to real process analysis. We are still developing the ideas of process viewpoints and plan to carry out the first experimental trials of these ideas in mid-1995. Our intention is firstly to carry out paper studies using process data derived from ethnographic analyses then to apply them in an industrial context in studies of the requirements engineering process.

5 Acknowledgements

This research has been partially supported by the European Community's Framework Programme of IT research in the REAIMS project (8649).

References

[1] Heineman, G.T., *et al.*, "Emerging technologies that support a software process life cycle". *IBM Sys. J.*, 1994. **33**(3): p. 501-29.

[2] Rodden, T.A., *et al.* "Process Modelling and Development Practice". in *EWSPT'94*. 1994. Villard de Lans, France.

[3] Sommerville, I. and T. Rodden, "Social and Organisational Influences on Software Process Evolution", in *Trends in Software Processes,* A. Fuggetta and A. Wolf, Editors. 1995, John Wiley and Sons: New York.

[4] Basili, V.R. and H.D. Rombach, "The TAME project: Towards Improvement-Oriented Software Environments". *IEEE Trans. on Software Eng.*, 1988. **14**(6): p. 758-773.

[5] Kotonya, G. and I. Sommerville, "Viewpoints for requirements definition". *BCS/IEE Software Eng. J.*, 1992. **7**(6): p. 375-87.

[6] Finkelstein, A., *et al.*, "Viewpoints: A Framework for Integrating Multiple Perspectives in System Development". *Int. J. of Software Engineering and Knowledge Engineering*, 1992. **2**(1): p. 31-58.

[7] Easterbrook, S. "Domain Modelling with Hierarchies of Alternative Viewpoints". in *RE'93*. 1993. San Diego, USA.

Process-Based Software Risk Assessment

Alfred Bröckers

AG Software Engineering
Department of Computer Science
University of Kaiserslautern
D-67653 Kaiserslautern, Germany
Email: broecker@informatik.uni-kl.de

Abstract. Analyzing software process models to predict the behavior of software processes helps in planning and enacting software projects. Since software process models can capture the key information that is necessary to assess process-related risks, this paper discusses how approaches for software process analysis may be applied to software risk assessment. A characterization scheme for process analysis approaches is stated based on a set of necessary risk assessment requirements. Existing analysis approaches are evaluated with respect to the characterization scheme. Proceeding from this evaluation, an approach to software process analysis is proposed that is specifically tailored to software risk assessment. This analysis approach takes advantage of a wide range of information by integrating empirically validated models. It is also shown how this approach fits into the context of the MVP (multi-view processes) project at the University of Kaiserslautern.

Keywords

Software process analysis, survey of software process analysis approaches, process-based risk assessment, risk analysis, MVP project.

1 Introduction

Explicitly modeling the software development process is an important step in evolving the current state of software development practice from an applied art to an engineering discipline. As Lonchamp [17] points out, software process modeling aims at expressing, guiding, and analyzing software processes. Most current research in the software process area focuses on modeling and guiding software projects. Far less work is done on software process analysis that is based on models of software development processes. Among the very few approaches for software process model analysis there is Abdel-Hamid's approach that aims at investigating general phenomena that can be observed in a wide range of software processes [1, 2]. Approaches such as SLICS [15] deal with estimating certain project parameters throughout the software development process. Other approaches such as MELMAC [9] deal with properties that are meaningful for planning and managing software processes.

    Software process models can capture the key information that is necessary to perform software risk assessments. In addition, software process models possess more expressive power than representations that underly traditional risk assessment techniques. Therefore, one might assume that approaches for software process model analysis can be used to identify and analyze process-related risks. According to Boehm [3], the goal of risk assessment is to identify risks and to analyze each identified risk in terms of loss magnitude and loss probability. Therefore, risk assessment requires the ability to obtain results concerning quantitative outcomes such as time or effort and their occurrence probabilities. It appears that no existing software process analysis approach produces this kind of results. However, some existing approaches are capable of identifying possible risks.

The paper evaluates existing approaches to software process analysis with respect to their ability of supporting software risk assessment. It introduces a novel approach to the analysis of software process models that is tailored to software risk assessment. Section 2 gives a short introduction to the terminology that is used throughout the paper. Section 3 shows how software process analysis is related to established risk assessment practices and states some inherent requirements for risk assessment support via software process model analysis. Section 4 introduces a classification scheme that is based on these requirements and applies it to existing approaches for software process analysis. Section 5 shows how an existing software process modeling approach, namely the Multi-View Processes(MVP) approach [6], can be extended so that it can be used as a basis for software risk assessment. Section 5 further demonstrates the necessary extensions, and provides a conceptual architecture of a support system for assessing process-related risks. Section 6 summarizes the results and states some future research directions.

2 Terminology

After some terminological confusion was observed in the software process modeling community, some work was done on defining the most commonly used terms [10, 17]. Wherever possible, the paper relies on those definitions. Since the focus of the paper is on a specific application domain of software process model analysis, these general definition efforts do not provide a sufficient level of detail. Thus Sect. 2 introduces some terms that are important for understanding the paper. Furthermore, the basic concepts and terms of software risk assessment are described here.

Software process: Lonchamp [17] defines a software process as

> *A set of partially ordered process steps, with a set of related artifacts, human and computerized resources, organizational structures and constraints, intended to produce and maintain the requested software deliverables.*

In this definition a process step is a subprocess aimed at achieving a subgoal. Thus the only difference between a process and a process step is, that the latter does not deal with the overall project goal but with one or more subgoals. Therefore, in the paper we will refer to both as software processes or processes for short.

Process model: A software process model, or process model for short, describes how a class of processes is expected to be performed. A process model represents one or more aspects. As Lonchamp points out, these aspects consist of structure, resource consumption, development artifacts, and constraints. In addition, a process model can describe goals and effects of a process on its own properties and on the status of development artifacts and resources.

Product model: The main goal of a software process is the development or maintenance of a deliverable software product. Abstractions that show how a class of software products should look are called product models. In addition to the final product, abstractions of by-products, temporary artifacts, and parts of a software product's documentation are similarly called product models. Thus models of each artifact that is relevant throughout the course of software development are called product models.

Resource model: A resource model is an abstraction of resources that are necessary for performing a software process. Resource models describe personnel resources such as test engineers or a group of system analysts, and material resources such as tools and workstations.

Qualitative versus quantitative models: On the one hand, process, product, and resource models can describe single processes, products, and resources, or limited sets of identifiable entities and their qualitative characteristics. These kinds of models are called *qualitative models*. An example is a model for an acceptance test process which identifies the integrated system and the requirement specification as input products and a test report as the only output product. This model can be instantiated into multiple, identifiable acceptance test processes.

On the other hand, aspects of processes, products, resources, and their relations can be defined quantitatively on a ratio scale. Although these quantitative representations can be used to describe single properties of processes, products, and resources, they themselves do not describe instantiations of these entities. An example is a model of project progress that is based on the amount of completed source code.[1] Another example is a process model that relates the requirement specification complexity (in function points) to the overall acceptance test effort. These kind of models are called *quantitative models*.

Qualitative and quantitative models are not mutually exclusive within a particular software process modeling approach. They can be combined to model entities and quantitative knowledge of software processes.

Stochastic versus non-stochastic models: Fenton [11] points out that most of the attributes of processes, products, and resources are stochastic rather than deterministic. As a consequence, process, product, and resource models can include their stochastic behavior. In terms of qualitative models, this means that the occurrence probability of an event can be included in a model. Stochastic, quantitative models include stochastic information such as mean, probabilistic distribution, skewness, etc. As an example, the process model for the acceptance test can include the probability for a system to pass the test successfully.

---

[1] The question of whether this model makes sense or not is not covered here.

Software process analysis: Lonchamp [17] points out that when analyzing software process models:

> *the description of the actual or desired process is studied through more or less formal techniques (...) for a deeper understanding, comparisons, improvement, impact analysis, or forecasting.*

Desired or actual processes are typically represented by sets of process, product, and resource models.

Barry Boehm and Robert N. Charette independently tried to clarify the basic concepts of software risk management at the same time [3, 7]. Unfortunately their terminological definitions appear to conflict. The remainder of this section introduces the most important concepts and terms of software risk management based on Boehm's definitions.

Boehm [3] defines risk management as

> *... an emerging discipline whose objectives are to identify, address, and eliminate software risk items before they become either threats to successful operation or major sources of rework.*

Boehm divides software risk management into risk assessment and risk control, each of which is divided into three subsidiary steps (Fig. 1).

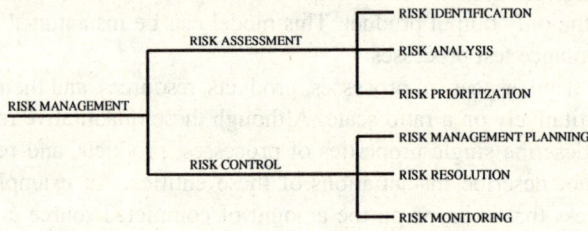

Fig. 1. Software risk management steps[3]

Risk control deals with managing known and assessed risks during software development. Since the paper concentrates on software risk assessment, a further description of software risk control is not provided here. For a detailed description see [3].

Risk assessment deals with identifying, analyzing, and prioritizing software risk items as follows:

Risk identification: To tackle the risks of a specific software project, it is clearly necessary to identify the risks one could face. Widely applied techniques include risk identification checklists, assumption analysis, and decomposition. The application of risk identification checklists involves the investigation of presence or absence of risks that have been shown to be relevant. Examples of risk identification checklist entries are the familiarity of the developers with the application domain and the experience of the designers with the design technique. Another frequently applied technique is assumption analysis that deals with the identification of optimistic assumptions about the performance of a project. Decomposing the overall project into processes and their dependencies can be a valuable help in identifying risks that are caused by unidentified dependencies between development activities. Often these dependencies are implicit and project-specific. For example, they can be caused by conflicting needs for resources such as personnel. A frequently applied technique is the use of PERT charts to represent the processes and known dependencies.

Risk analysis: Risk analysis deals with determining the extent to which a software risk item can harm a software project. That means the magnitude and the probability of loss or gain are identified. Losses or gains in software projects materialize as alternate project outcomes. The *risk exposure* is calculated as the sum of the products of alternate project outcomes and their respective probabilities. Frequently applied risk analysis techniques include decision tree analysis, network analysis, and cost-risk analysis. The decision tree is a concept that combines two or more planning alternatives, their respective alternate outcomes, and their probabilities into a single tree structure.

Network analysis aims at the analysis of risks that are caused by deviations from ideal activity durations, by resource consumptions, or by uncertainties about decision results. Classical network analysis involves the computation of schedules based on a description of activity dependencies represented as an directed, acyclic graph. Nodes of this graph represent activities and are attributed with the lower and upper bounds of the expected activity duration.[2] More sophisticated network analysis approaches include the probabilistic distributions of the duration of each activity and involve multiple simulations to compute probability distributions of software process outcomes such as overall project duration. A widely accepted weakness of this kind of network analysis is its lack of consideration of rework or iterations in software processes. Therefore, some network analysis approaches such as GERT allow for inclusion of alternate paths and iterations. Another problem is that the dependencies, no matter if explicit or implicit, must be known and modeled explicitly. Williams [21] presents a more advanced technique that in addition to activity durations can include uncertainties concerning resource requirements of activities, resource availability, and even the network structure. However, such approaches are very general and do not reflect specifics of software processes.

An alternate approach to evaluating the possible outcomes of software projects is the use of cost models such as COCOMO [4]. These models can be used to calculate effort and overall project duration based on known or estimated project parameters such as program size. Alternate outcomes associated with risks are achieved by varying these parameters. Computing the probabilistic distribution of the outcomes is possible by performing multiple simulations based on estimated or empirically validated knowledge of probabilistic distributions of the independent model parameters such as estimated product size (measured in KLOC.) This kind of analysis aims at risks that are caused by uncertainties about these parameters. The problem with this cost-model-based risk analysis is that the set of parameters which can be varied is restricted to a comparatively

---

[2] There are versions of network analysis where the edges are attributed with the duration bounds. This has no effect on the descriptive and analytic power.

small set.[3] Another problem is the generality of cost models. Project specifics cannot be covered to a great extent.

Risk prioritization: Seldom is it possible to deal with all of the risks which a software project could face. To decide which subset of the identified and assessed software risk items should be dealt with, it is necessary to identify the most harmful risk items. Therefore, risk prioritization produces a prioritized list of software risk items.

3 Process-based risk assessment

Boehm presents a list with the primary sources of software project risks in [5]. Figure 3 shows an excerpt from this list showing the top five risk items. Number two of this list is "unrealistic schedule and budget." "Unrealistic" implies that these risks are not caused by the software process performed and its environmental constraints. These risks are caused by invalid assumptions about the overall process, process elements (processes, products, or resources), and their interaction. I will refer to risks that are related to the overall process, process elements, and their interaction as *process-related risks*. Currently, stochastic network analysis techniques are used to identify and analyze process-related risks. The problem with network analysis techniques is that the only input information consists of the stochastic distribution of the duration of each activity, resource availability, and successor-predecessor relationships. Another limitation is that activity duration is treated as an independent variable. Interactions between activities and dependencies of the process performance upon process, product, and resource attributes are ignored. For example, the complexity of a requirement document can influence the acceptance test effort strongly. Another example is the impact a software designer's experience may have on product quality and design effort. Since activity networks cannot represent this kind of information, network analysis cannot process it. In contrast to activity networks, software process models can describe this kind of information. Consequently, software process analysis is a candidate for investigation of process-related risks. The use of software process analysis for identification and analysis of process-related risks is referred to as *process-based risk assessment*. Process-based risk assessment promises more realistic results than traditional risk assessment techniques by including more realistic models. These models can represent a variety of risk driver information, allowing for tracing from risk analysis results back to the causes.

The remainder of this section is structured as follows. First Sect. 3.1 shows how process-based risk assessment can support or even replace established risk assessment practices. Section 3.2 identifies some risk assessment-related requirements for software process analysis approaches. In Sect. 4 a classification scheme is proposed that is based on these requirements.

3.1 Relationship to Traditional Risk Assessment Techniques

Process-based risk assessment aims at identifying and analyzing process-related risks. The following discussion of the relationship of process-based risk assessment to cur-

---

[3] This restriction is explicitly done to exclude interrelated parameters.

1. Personnel shortfalls
2. Unrealistic schedules and budget
3. Developing the wrong software functions
4. Developing the wrong user interface
5. Gold plating

Fig. 2. The top five software risk items [3]

rently applied techniques is divided according to the two risk assessment steps risk identification and risk analysis.

Process Analysis Support for Risk Identification. One of the essential features of software process analysis is that it provides for the investigation of alternate project characteristics. Among these alternate project characteristics are the presence or absence of certain risky situations. Thus process analysis can be used to investigate whether a planned software project possesses the process-related risk items given on a checklist. For example, all possible process states according to the project plan can be computed. Then this set of states can be searched for problematic situations that are associated with a software risk item on a checklist. Risk identification checklists can be built directly into supporting software process analysis tools.

Assumption analysis can be supported by comparing the assumptions about the performance of a software project with the results of software process analysis. For example, process model analysis may be used to identify alternate paths through a software project that can be compared to an assumed schedule.

By modeling software processes, software projects are decomposed into process, product, and resource models. Software process model analysis can be used to identify hidden dependencies and their implications by combining and integrating all information about explicit dependencies of processes, products, and resources. Thus decomposition as a risk identification technique can be supported directly.

Process Analysis Support for Risk Analysis. Software process analysis techniques can be used to compute the quantitative outcome of different ways of performing a software project. In addition, they can provide information about the probabilities of each performance. For example, process analysis can be used to compute the set of possible paths through a software project and their probabilities. The outcome and probability values provided by software process analysis can be used as input to decision tree analysis.

Process analysis can also be used to compute implications of dependencies between processes, products, and resources on project outcomes, such as schedule and overall cost. Thus, depending on the expressive power of the underlying modeling formalism, network analysis techniques can be replaced. In contrast to activity network analysis, process analysis can include iterations and is not limited to investigating the effects of obvious dependencies.

3.2 Requirements for Process Analysis Support

The suitability of a software process analysis approach for supporting the risk analysis techniques presented above depends strongly on the extent to which the approach provides the required result type, and the extent to which it includes information on potential risk drivers, such as hidden activity dependencies or resource assignments. While the former is obvious, the latter results from the fact that the interesting results arise from varying the risk drivers. The following list of requirements is divided into requirements that are related to the analysis results and requirements that are related to the expressive power of the underlying software process modeling language.

Requirements concerning the result types:

Requirement 1 (Alternate project outcomes): An approach to process-based risk assessment must provide for the computation of alternate project outcomes. Among these outcomes are desirable or tolerable outcomes and outcomes that represent the materialized risks. These outcomes can either be represented by global project attributes or by attributes of intermediate project situations. Among the global project attributes that are relevant to software risk assessment, some attributes such as overall project duration and cost can be estimated using cost and effort models. Other examples are attributes like average and maximum need for personnel, average personnel work load, etc. Some process-related risks are not related to global process attributes but to undesirable, intermediate project situations. Examples are bottlenecks in a software process or undesirable resource assignments. Therefore, in addition to the global project attributes, the occurrence of intermediate project situations and project paths are software project outcomes that are relevant to risk assessment. Sequences of project states and activities are especially of great interest. Examples of risk items involving project states and activity sequences are deadlock situations, race conditions, or bottlenecks.

Requirement 2 (Stochastic results): An approach to process-based risk assessment must provide for computing the occurrence probability for each alternate outcome. In combination with requirement 1, this allows for computation of risk exposure values. On the one hand, the results may be given as probability values for discrete outcomes of a software project. On the other hand, the probabilities can be represented by the continuous distribution of an outcome. Examples for these kinds of stochastic results are the probability of a deadlock situation, probabilities of rework situations, and the probabilistic distribution of the overall project effort. Predicting the probabilities of alternate outcomes is a necessary prerequisite of risk analysis.

Requirements concerning the risk driver representation:

Requirement 3 (Inclusion of qualitative models): Risks of a software project are often caused by implicit or unknown dependencies of entities, such as processes, products, and resources, upon each other. For example, the fact that an activity changes a product while another activity consumes this product can imply their sequential performance. Another example is a resource conflict of activities that implies their mutual exclusive performance, which in turn may lengthen the overall project duration. Therefore, an approach to process-based risk assessment must include qualitative models.

Requirement 4 (Inclusion of quantitative models): Much information concerning the behavior and the results of software processes can only be given as quantitative models. These models can either represent global attributes of software projects, such as project effort or overall project duration, or describe characteristics of single entities within a software project, such as processes, products, or resources. Thus an approach to process-based risk assessment must include information that is represented as quantitative models.

Quantitative models show the dependency of a dependent variable (the characteristic of interest) on zero or more independent variables. An example is a model that relates the expected test effort of a given test process to the structural complexity of a component.

Requirement 5 (Inclusion of stochastic models): Process-based risks are often caused by deviations of process, product, and resource attributes from their expected values. Therefore, an approach to process-based risk assessment must include information about these deviations as stochastic models. In addition, the inclusion of stochastic models is a prerequisite for the computation of non-simplistic stochastic results.

Requirement 6 (Reuse of models): The validity of the results of process-based risk assessment depends highly on the reliability of the input information. Modeling software processes at the level of detail that is required to perform a comprehensive risk assessment is an expensive and ambitious task. Therefore, an approach to process-based risk assessment must provide for inexpensive reuse of process, product, and resource models.

4   A Comparison of Existing Analysis Approaches

This section presents a classification scheme that is based on the requirements stated in Sect. 3.2. The classification scheme is then used to evaluate the applicability of existing software process analysis approaches to process-based risk assessment. Section 4.1 presents the classification scheme. Section 4.2 briefly introduces some of the better-known approaches to software process model analysis and classifies them according to the scheme.

4.1   Classification Scheme

Facet 1 (Alternate project outcomes): This facet is related to requirement 1 and specifies whether an analysis approach has the ability to compute alternate project outcomes. Possible values are *none* and *supported.*

Facet 2 (Stochastic results): This facet is related to requirement 2 and specifies whether an analysis approach provides for computation of stochastic results. Possible values are *none* and *supported.*

Facet 3 (Inclusion of qualitative models): This facet specifies the extent to which a process analysis approach takes advantage of qualitative models. An analysis approach can either possess a fixed set of qualitative models of processes, products, or resources, or allow for the flexible definition of these models. In the first case, the only way to adapt the models to specific project needs is to define the number of entities, such as the number of system analysts on a project. Additional important information is the significance of alternate behavior. For example, a process model can include information concerning the probabilities of two or more alternate, qualitative process results. Therefore, possible values for this facet are *none, fixed non-stochastic, variable non-stochastic, fixed stochastic, variable stochastic.*

Facet 4 (Inclusion of quantitative models): This facet specifies the extent to which a process analysis approach takes advantage of quantitative models. An analysis approach can either have an underlying fixed set of quantitative models that can be influenced by parameters, or allow for the flexible definition of quantitative models.

In addition to representing single values for certain characteristics, quantitative models can include information about their probabilistic distribution. That means that an analysis approach can either be restricted to non-stochastic models or can include stochastic quantitative models. Possible values for this facet are *none, fixed non-stochastic, variable non-stochastic, fixed stochastic, variable stochastic.*

Facet 5 (Reuse of models): Process analysis is just one of several objectives for modeling software processes. The simplifying assumption of this facet is that the coverage of objectives is related to the availability of models for reuse. Therefore, the objectives for modeling software processes are used as values in our scheme. Possible values are *understanding, estimating, planning, guidance,* and *controlling.*

Fig. 5 summarizes the classification facets and their possible values.

Facet	Possible values
1 alternate project outcomes	none, supported
2 stochastic results	none, supported
3 inclusion of qualitative models	none, fixed stochastic, fixed non-stoch., variable stochastic, variable non-stoch.
4 inclusion of quantitative models	none, fixed stochastic, fixed non-stoch., variable stochastic, variable non-stoch.
5 reuse of models	understanding, estimating, planning guidance, controlling

Fig. 3. Classification scheme

4.2 Classification

This section shortly introduces some well-known approaches to software process model analysis and classifies them according to the classification scheme of Sect. 4.1.

Statemate

Kellner and Hansen [14] present an approach to modeling software processes from three viewpoints (behavior, structure, and functionality). These three viewpoints are modeled using state charts, module charts, and activity charts, which were developed as a specification and analysis technique for reactive real-time systems [13].

Kellner and Hansen utilize the Statemate system which supports state charts, module charts, and activity charts directly [14]. In addition to providing modeling facilities, it also provides functionality for analyzing several system properties. These include completeness, consistency, deadlocks, and non-determinism. The Statemate system possesses mature simulation functionality and is capable of searching for reachable system states.

Using Statemate one is able to produce alternate project outcomes (facet 1: supported). Statemate does not give any information concerning the probabilities of alternate project outcomes (facet 2: none). With respect to the characterization scheme, state charts, module charts, and activity charts can be regarded as variable, non-stochastic, qualitative models of processes, products, and resources (facet 3: variable non-stochastic, facet 4: none). Statemate models of processes, products, and resources are built for understanding and analysis purposes (facet 5: understanding, analysis).

MELMAC

Gruhn and Deiters introduce an integrated approach for modeling, analysis, and guidance of software projects [9]. They propose a high-level Petri net extension, called FUNSOFT nets, for modeling processes, products, and resources. The MEL-MAC system supports modeling, analysis, and enactment of software processes that are represented as FUNSOFT nets. In addition to checking several consistency rules, MELMAC is capable of simulating software projects. Furthermore, it allows for the analysis of reachable process states and for the detection of several properties such as deadlocks and race conditions.

MELMAC cannot compute different alternate project outcomes directly. The user is able to obtain the required alternatives by performing multiple simulations and reachability analyses (facet 1: supported). Although the MELMAC simulation facility considers probabilistic information as an input, it is not capable of producing results concerning the probabilities of the obtained project alternative (facet 2: none). The structure, nodes, and tokens of a FUNSOFT net represent processes, products, and resources of a project. FUNSOFT nets allow for probability specification of alternate outcomes of activities. Thus the FUNSOFT net representations of processes can be regarded as variable, stochastic, qualitative process models. Models for products and resources are variable, qualitative, non-stochastic models (facet 3: variable, stochastic (processes)). Furthermore each of the FUNSOFT net entities is extended by a fixed set of attributes that represent additional quantitative information such as expected activity duration. Thus with respect to the characterization scheme, these attributes can be viewed as a fixed set of non-stochastic, quantitative models of processes, products, and resources (facet 4: fixed non-stochastic). FUNSOFT representations of processes are built for understanding, analysis, and enaction purposes. Throughout a software project the MELMAC execution mechanism offers a set of possible activities to the developers who in turn inform the execution mechanism about the activities actually performed (facet 5: understanding, analysis, guidance).

DesignNets

Liu and Horowitz [16] propose a hybrid language called DesignNets for the representation of software processes. DesignNets are a combination of and/or graphs and Petri nets. And/or graphs are used to model the structure of software processes and products, while Petri nets are used to model product and control flow. Liu and Horowitz define some properties that are relevant to project management and provide appropriate analysis algorithms. Among these properties are internal consistency of the representation, contribution of each process to the deliverable software product, and overall project cost. Analysis results concerning alternate project outcomes are restricted to computing the overall costs of alternate representations (facet 1: supported). Results concerning the probabilities cannot be achieved (facet 2: none). DesignNets can be described as qualitative models of processes, products, and resources. Models of processes are augmented with a fixed set of quantitative models such as a model for effort (facet 3: variable non-stochastic, facet 4: fixed non-stochastic). DesignNet models are mainly built for planning, estimating, and guiding software projects (facet 5: planning, estimating, guidance).

Articulator

Mi and Scacchi [19] propose a software process modeling and simulation system called Articulator. The Articulator is a knowledge-based approach and uses a meta model of software processes. It supports the analysis of software processes by providing software process simulation functionality. User-defined queries can be applied to simulation histories. Furthermore, it is possible to perform what-if analyses by starting simulation from a given project state. The Articulator is not capable of achieving results concerning alternate project outcomes directly. However, a user can achieve these results by performing multiple simulations and analyzing the simulated project histories using identical (or similar) queries (facet 1: supported). The obtainable results are non-stochastic (facet 2: none). The range of possible process, product, and resource models is defined by the Articulator's meta model and can be regarded as variable, non-stochastic, qualitative models (facet 3: variable non-stochastic, facet 4: none). Articulator models are not built for analysis in isolation. The main focus of these models is understanding and guidance (facet 5: understanding, guidance).

System Dynamics

Abdel-Hamid and Madnick [2] divide a software project into four subsystems, namely human resource management, controlling, planning, and software production. Each of these models and their interfaces are modeled using System Dynamics models [12]. The main goal of this approach is the investigation of general phenomena in software projects such as Brook's law [2] or the effects of personnel flow between projects [1]. The required analysis results are obtained by simulating the System Dynamics representation. The results that are achieved by the proposed simulation do not involve alternate project outcomes directly. Alternate results can be achieved by simulating alternate project representations (facet 1: supported). Results do not include information concerning probabilities (facet 2: none). However, the set of underlying System Dynamics models remains fixed (although the parameters can be changed) within this approach. The System Dynamics models used in this approach are fixed, non-stochastic, quantitative models of processes, products, and resources and were solely built for simulation purposes (facet 3: none, facet 4: fixed non-stochastic, facet 5: analysis).

SLICS

Similarly to the System Dynamics approach presented above, Lin and Levary [15] use System Dynamics models to represent processes, products, and resources. They introduce the SLICS system (system life cycle simulator) that supports the simulation of these models. In contrast to Abdel-Hamid's approach, Lin and Levary divide a software project into three subsystems, namely staffing, product, and budget. The goal of this approach is the continuous estimation of the effects of management decisions and assumptions on several project variables such as the needed staff profile, overall cost, and schedule of the project. Continuous means that these variables can be reestimated throughout the project whenever an uncertain input variable can be replaced by collected data.

With respect to the facets, the two approaches that involve System Dynamics models are similar. Thus varying the project parameters provides for alternate project outcomes (facet 1: supported). The results that are achieved by SLICS can be viewed as non-stochastic results (facet 2: none). The underlying set of non-stochastic, quantitative models is fixed with the possibility of changing some parameters (facet 3: none, facet 4: fixed non-stochastic). Within the SLICS approach, system dynamics models are built solely for estimation purposes (facet 5: estimating).

ProNet

Christie [8] uses a graphical notation called ProNet as a modeling language. ProNet is an extension of the entity-relationship model and provides entity types for processes, products, resources, etc. A set of predefined relationship types is used to model relationships between the entities such as data flow. Throughout a software project an execution mechanism provides guidance to the developers by offering a set of possible activities. The software developers inform the execution mechanism about the activities actually performed. After project termination, the models of processes, products, and resources are simulated and checked against the actual project performance to detect deviations from the initially modeled behavior. The analysis results do not involve alternate project outcomes since the performed simulation tries to match a single, actually performed project history (facet 1: none). Stochastic information is not obtainable either (facet 2: none). ProNet models can be viewed as variable, qualitative, non-stochastic models of processes, products, and resources (facet 3: variable non-stochastic, facet 4: none). Building ProNet models aims at providing guidance for developers and at performing post-mortem process conformance analyses (facet 5: guidance, controlling).

Figure 4 summarizes the results of the characterization. None of the compared approaches fulfills all of the risk assessment requirements set out in the paper. With the exception of the ProNet approach, all approaches are capable of computing alternate project outcomes (facet 1), but none support stochastic results (facet 2). Thus the compared approaches do not provide full risk assessment support. The MELMAC approach fulfills (at least partially) all requirements that are related to risk identification (facets 1, 3 – 5). Most of the remaining approaches (Statemate, DesignNet, Articulator, System Dynamics, and SLICS) are restricted to identification of a smaller set of risks since they either focus strongly on non-stochastic qualitative models (Statemate, DesignNet, Articulator), or on non-stochastic quantitative models (System Dynamics, SLICS). The ProNet approach deals with analyzing past projects. It does not support predicting outcomes of future projects. Therefore, the ProNet approach does not support process-based risk assessment.

	Statemate	MELMAC	DesignNet	Articulator	System Dynamics	SLICS	ProNet
Facet 1 Alternate outcomes	supported	supported	supported	supported	supported	supported	none
Facet 2 Stochastic results	none	none	none	none	none	none	none
Facet 3 Qualitative models	variable non-stoch.	variable stochastic (processes)	variable non-stoch.	variable non-stoch.	none	none	variable non-stoch.
Facet 4 Quantitative models	none	fixed non-stoch.	fixed non-stoch.	none	fixed non-stoch.	fixed non-stoch.	none
Facet 5 Reuse of models	understanding analysis	understanding analysis guidance	planning estimating guiding	understanding guidance	analysis	estimating	guidance controlling

Fig. 4. Characterization of the analysis approaches

5 Process-based Risk Assessment in the Context of MVP

The main goal of the MVP project at the University of Kaiserslautern is the development of a process-centered development environment that supports project-specific planning and management activities and provides guidance to developers throughout the software project [18]. This environment is called MVP–S (MVP system). Within the MVP project the software process modeling language MVP–L was developed to support the integration of multiple views of software processes and the integration of measurement [20]. This section introduces the basic ideas behind a MVP-based risk assessment approach. Section 5.1 briefly introduces the main concepts of MVP–L and classifies them according to facets 3 – 6 from Section 4.1. Section 5.2 shows how additional information may be added to MVP–L models to provide a sufficient basis for process-based risk assessment. Section 5.3 proposes a high-level architecture of a MVP-based system that will support process-based risk assessment.

5.1 The Language MVP-L

The main elements of the MVP–L language are models of processes, products and resources that can be combined and instantiated into a project plan. The distinction between processes, products, and resources is made explicit by having a model type for each of these groups. Each of these models can be subsequently refined into a set of models of the same type.

Attributes can be attached to process, product, and resource models. These attributes are themselves defined using attribute models. An attribute model describes an attribute's range and specifies the events that lead to a change in the attribute's value. The set of attributes of a process, product, or resource model and their corresponding attribute models completely define the possible behavior of these objects (while their actual behavior depends on the interaction of processes, products, and resources). Furthermore, attribute models can be used to integrate data collection tools that collect data either automatically or by requesting data from human software developers. The motivation for this integration possibility is twofold. First, it allows MVP–S to control and partially automate the data collection activities of a measurement program. Second, any decision process is regarded as a data collection effort that can either be done by a human or by an automated tool. For example, a developer is asked via a form if a product successfully passed a review process or if further rework is to be done. Figure 5 shows an example of a process attribute model for effort that involves data collection. Whenever a process possessing such an attribute is terminated, a data collection tool is invoked that collects the effort data from the software developer who performed the process.

```
process_attribute_model effort () is
   —— Attribute model for development effort (1.00 means 1 hour)
   ——

   attribute_type real;
   attribute_manipulation
           user_triggered
                  when complete
                        and others —> effort := request(effort—collection)
                                              in [0.30 .. 1000.00];
end process_attribute_model effort
```

Fig. 5. MVP–L attribute example

Process models are the integrating elements within MVP–L. Process models describe which products are to be consumed, produced, or changed by a process' execution. Furthermore, a process model specifies entry criteria as boolean conditions that define the states of the world required to start executing a process. Similarly, exit criteria denote the states expected upon termination. Process models contain a specification of the resources that are necessary to execute a process. The execution mechanism of MVP–L is event-based. That means a developer initiates changes by sending an event to a process model. This event is further propagated through products, processes, and resources leading to changes of attributes, and thus to a new overall project state.

Although MVP–L has an underlying textual notation, the remainder of this paper is restricted to MVP–L's graphical notation. As an example, Fig. 6 shows a process, two products, and a resource. Processes are represented by rectangles. A rhombus that is connected to the left of a process denotes the process' entry criteria, while a rhombus connected to the right represents its exit criteria. Products are depicted as circles. An arrow from a product to a process object means that this product is consumed by the process during execution. An arrow from a process to a product means that the process is responsible for producing the product. Object identifiers are separated from their type identifiers (model identifiers) by a colon and are placed above their graphical object (process or product). Resources are not depicted as graphical objects. They are solely represented by their object identifier followed by a colon and the resource model identifier. A resource is graphically attached to a process by a double line between the resource identifier and the rectangle that represents the process.

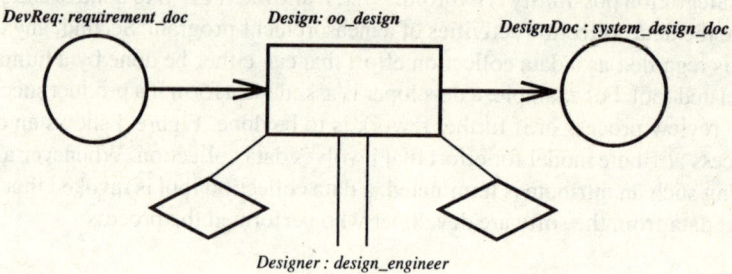

DevReq: requirement_doc   *Design: oo_design*   *DesignDoc : system_design_doc*

Designer : design_engineer

Fig. 6. Graphical notation for MVP–L

One basic idea of MVP–L is to provide language concepts that are meaningful in the application domain of software development. Thus it provides model types for each naturally identifiable class of entities relevant in a software development process. Consequently, with respect to the classification scheme from Sect. 4.1, MVP–L models have to be regarded as non-stochastic, qualitative models of processes, products, and resources. However, the language allows for specifying real-valued attributes. MVP–L does not provide any language concept for specifying how to obtain concrete values. The specification of a data collection tool cannot be regarded as a quantitative model. Thus MVP–L is neither capable of representing stochastic behavior nor does it provide concepts that are suitable for representing quantitative models.

5.2   Providing the Required Information

The preceding section showed that MVP–L lacks the expressive power necessary to perform a comprehensive, process-based risk assessment. However, MVP–L provides the possibility for integrating data collection tools via attribute models. This section shows how stochastic quantitative models are integrated into MVP–L models following the same integration strategy. For risk assessment purposes, we can link any call to a data collection tool to a quantitative model that specifies the expected probabilistic distribution of return values. This provides for sufficient inclusion of quantitative models and the possibility of adding stochastic information to MVP–L models.

MVP–L provides 5 different basic attribute types, namely integer, real, string, boolean, and enumerated. Each of these attribute types requires a different kind of distribution specification. The return value distribution for an enumerated type can be specified by providing a probability value for each of the qualitative values of the enumerated type. As an example, product model "module_design" in Fig. 7 possesses an enumerated type attribute. This attribute is bound to a distribution function that provides a probability value for values 'A', 'B', 'C', and 'D.' A return value distribution for a real type can be specified by a distribution curve, or by giving mean, variance, and a

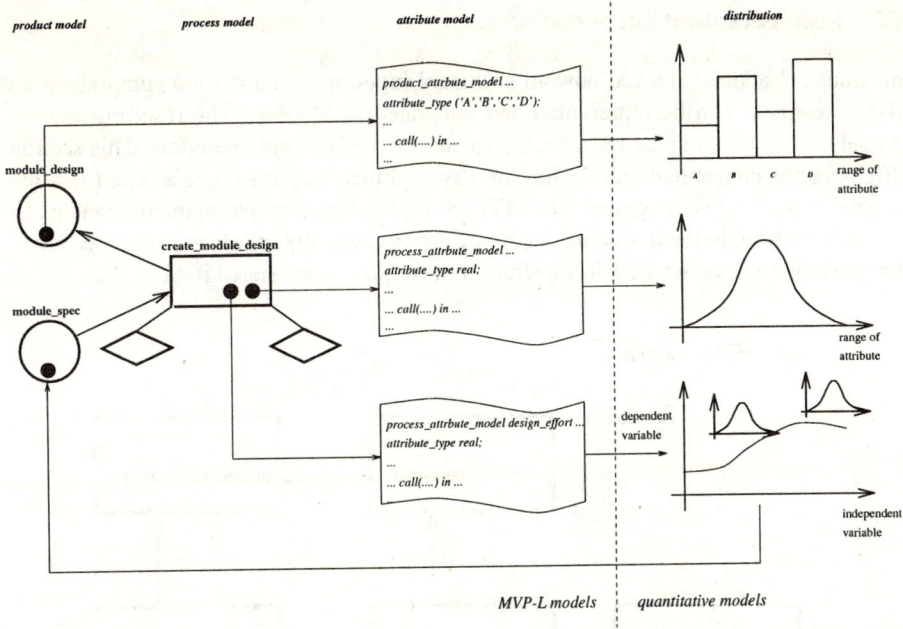

Fig. 7. Linkage of MVP–L models and quantitative models

distribution type, such as normal or Rayleigh distribution. The process attribute model in the middle of Fig. 7 is an example. The curve to the right of the attribute model indicates a normal distribution. The integration of quantitative models is not restricted to these kinds of simple models. Furthermore, it is possible to integrate quantitative models that describe how dependent variables are related to independent variables. As an example, the process attribute model "design_effort" at the bottom of Fig. 7 is linked to a quantitative model. The quantitative model describes the dependency of the module design effort on the module specification complexity. In Fig. 7 this dependency is represented as a curve to the right of the process attribute model. The small distribution curves indicate that a stochastic deviation from the curve has to be specified to reflect the stochastic nature of the dependency.

Guidance of a software process by MVP–S requires that most of the variables necessary to perform process-based risk assessment are represented by MVP–L attributes. Examples are process execution time or expected failure detection rate of a specific test process. Therefore, quantitative model building can be based on past experience naturally. Furthermore, measurement programs are performed within the context of MVP–L models. That means that any data collection is associated with a MVP–L attribute. The obtained quantitative models can be reused with the MVP–L models for risk assessment.

5.3 Risk Assessment Support

Sections 5.1 and 5.2 showed how all necessary input information for a comprehensive risk assessment can be represented and integrated in MVP–L. The resulting web of models is far too complex for a human to analyze with acceptable effort. This section discusses the conceptual architecture of a system that supports process-based risk assessment and that is integrated into MVP–S. Figure 8 shows the main components of this architecture. In the following, the basic ideas behind this architecture are presented. Furthermore it is shown how it implements the requirements stated in Sect. 3.2.

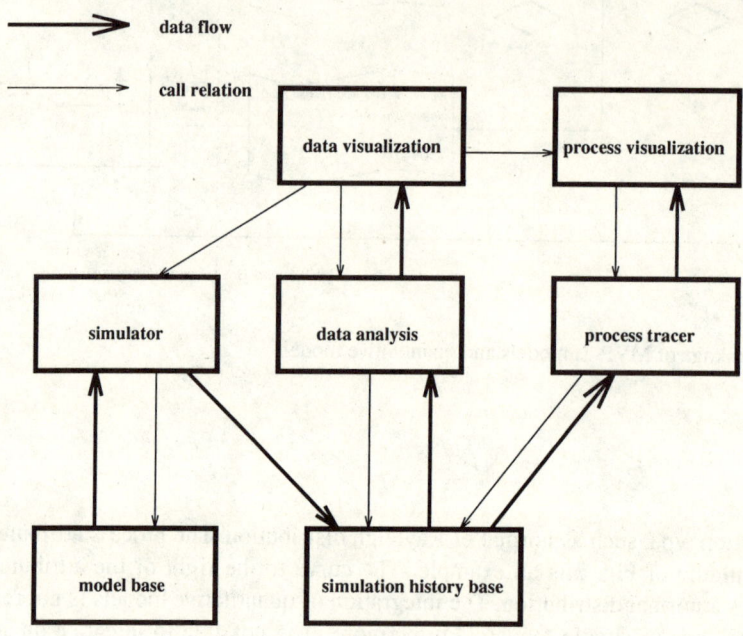

Fig. 8. Conceptual architecture of the analysis subsystem of MVP–S

As pointed out in Sect. 3.2, the risk analysis results are required to incorporate the probabilistic distribution of alternate project outcomes. The analysis technique must reflect this. The approach that is used here is a Monte Carlo simulation, since it promises to provide the required stochastic results. The idea is to perform multiple simulations and to aggregate simulation histories to the required distribution afterwards.

Multiple simulation: The component "simulator" performs multiple simulations based on a complete set of MVP–L models and associated quantitative models (component "model base"). The stochastic nature of the quantitative models ensures alternate project performances (requirement 1). Furthermore, it provides for realistic distribution of outcomes (requirement 2). Each alternate project performance is stored as a simulation history for data analysis (component "simulation history base").

Data analysis: Component "data analysis" computes the distribution of project attributes, such as schedule or effort, from the set of simulation histories (requirement 2). However, the approach is not restricted to a fixed set of relevant project attributes. A query mechanism allows for the computation of user-defined project attributes (component "data analysis").

Graphical presentation of results: The results of the data analysis are presented graphically. That means that in addition to the mean value (if possible, otherwise the median) of the project attribute of interest, the sample distribution is displayed graphically (component "data visualization").

Risk tracking: Risk assessment cannot be regarded as an isolated part of risk management. Furthermore, the results of risk assessment are essential inputs for risk control. It is not sufficient to identify the effects of risks. Undesired effects of risks have to be traced back to their causes. After selecting a sample value from the graphical distribution representation (component "data visualization"), the process visualization component is called. This component allows for tracing forward and backward through the associated simulation history (component "simulation history base" and component "process tracer") and for displaying the simulated project performance graphically (component "process visualization").

6 Conclusion

In the paper I introduced the analysis of formal software process models as a means for assessing process-related risks. Since this process-based risk assessment includes more realistic models and a wider range of risk driver representations than traditional techniques, it provides for more realistic results. I stated a classification scheme based on a set of inherent risk assessment requirements. This scheme was applied to some of the better-known approaches to software process analysis to evaluate their applicability to process-based risk assessment. The results of this classification were that the investigated approaches do not capture the knowledge of identifiable entities and quantitative knowledge to the extent necessary to support risk assessment. The approaches are not capable of providing stochastic results which is a prerequisite for full support of risk-assessment.

Proceeding from the observed weaknesses and strengths, the paper presented an approach to software process analysis that is tailored with respect to software risk assessment. This approach is being developed as a part of the MVP Project at the University of Kaiserslautern. It integrates quantitative, stochastic models into qualitative MVP–L models, and utilizes a Monte Carlo approach to provide for the required quantitative, stochastic results. The approach covers a wide range of process-related risks by supporting the analysis of user-defined aspects.

Currently the proposed analysis facilities are being built and integrated into the MVP system. Three major research efforts on process-based risk assessment are planned within the MVP project. First, more sophisticated techniques such as AI planning techniques can act as complementary techniques or as replacements for simulation.

Second, the applicability must be validated in industrial settings. A current measurement program provides for the process description and for quantitative models as a testbed for the first validation efforts. Third, we must investigate how much confidence we can have in the obtained results. We use stochastic models of processes, products, and resources to describe uncertainty, totally ignoring the fact that there is some uncertainty associated with the models themselves. However, this fact does not affect the benefits of process-based risk assessment. Carefully planning software projects should be based on as much information as is available, and process-based risk assessment is a step in this direction.

References

1. Tarek K. Abdel-Hamid. A multiproject perspective of single-project dynamics. *Journal of Systems and Software*, 22(3):151–165, September 1993.
2. Tarek K. Abdel-Hamid and Stuart E. Madnick. Lessons learned from modeling the dynamics of software development. *Communications of the ACM*, 32(12):1426–1438, December 1989.
3. Barry Boehm. *Software Risk Management*. IEEE Computer Society Press, 1989.
4. Barry W. Boehm. *Software Engineering Economics*. Advances in Computing Science and Technology. Prentice Hall, 1981.
5. Barry W. Boehm. Software risk management: Principles and practices. *IEEE Software*, 8:32–41, January 1991.
6. Alfred Bröckers, Christopher M. Lott, H. Dieter Rombach, and Martin Verlage. MVP Language Report. Technical Report 229/92, Fachbereich Informatik, Universität Kaiserslautern, December 1992.
7. Robert N. Charette. *Software Engineering Risk Analysis and Management*. McGraw-Hill, NY, 1989.
8. Alan M. Christie. Process-centered development environments: an exploration of issues. Technical Report CMU/SEI-93-TR-4, Software Engineering Institute, Carnegie Mellon University, Pittsburgh, PA 15213, June 1993.
9. Wolfgang Deiters and Volker Gruhn. The FUNSOFT net approach to software process management. Technical Report ISST 2 (ISSN 0943-1624), Institut für Software- und Systemtechnik, Fraunhofer-Gesellschaft e.V., 44227 Dortmund, February 1993.
10. Peter H. Feiler and Watts S. Humphrey. Software process development and enactment: concepts and definitions. Technical Report CMU/SEI-92-TR-04, Software Engineering Institute, Carnegie Mellon University, Pittsburgh, PA, September 1992.
11. Norman Fenton. Software measurement: A necessary scientific basis. *IEEE Transactions on Software Engineering*, 20(3):199–206, March 1994.
12. Michael R. Goodman. *Study Notes in System Dynamics*. Wright-Allen Press, Cambridge Massachusetts 02142, 1974.
13. David Harel, Hagi Lachover, Ammon Naamad, Amir Pnueli, Michal Politi, Rivi Sherman, Aharon Shtull-Trauring, and Mark Trakhtenbrot. Statemate: A working environment for the development of complex reactive systems. *IEEE Transactions on Software Engineering*, 16(4), April 1990.
14. Mark I. Kellner. Software process modeling: value and experience. In *SEI Technical Review*, pages 23–54. Software Engineering Institute, Pittsburgh, Pennsylvania 15213, 1989.
15. Chi Y. Lin and Reuven R. Levary. Computer–aided software development process design. *IEEE Transactions on Software Engineering*, 15(9):1025–1037, September 1989.

16. Lung-Chun Liu and Ellis Horowitz. A formal model for software project management. *IEEE Transactions on Software Engineering*, 15(10):1280–1293, October 1989.

17. Jaques Lonchamp. A structured conceptual and terminological framework for software process engineering. In *Proceedings of the 2nd International Conference on the Software Process*, pages 41–53. IEEE Computer Society Press, February 1993.

18. Christopher M. Lott. Process and measurement support in SEEs. *ACM SIGSOFT Software Engineering Notes*, 18(4):83–93, October 1993.

19. Peiwei Mi and Walt Scacchi. A knowledge-based environment for modeling and simulating software engineering processes. *IEEE Transactions on Knowledge and Data Engineering*, 2(3):283–294, September 1990.

20. H. Dieter Rombach, Alfred Bröckers, Christopher M. Lott, and Martin Verlage. Development environments for support of quality-oriented project plans (in German). *Softwaretechnik-Trends: Mitteilungen der GI-Fachgruppen 'Software-Engineering' und 'Requirements-Engineering'*, 13(3):1–8, August 1993.

21. T.M. Williams. Risk analysis using an embedded CPA package. *International Journal of Project Management*, 8(2):84–88, 1990.

The Use of Roles and Measurement
to Enact Project Plans in MVP-S

Christopher Lott,* Barbara Hoisl,** and H. Dieter Rombach

Research Group for Software Engineering
Department of Computer Science
University of Kaiserslautern
67653 Kaiserslautern, Germany
{lott, hoisl, rombach}@informatik.uni-kl.de

Abstract. Software development organizations are beginning to recognize that measurement is a prerequisite for systematic process improvement, and have started to measure their products and processes in order to understand, analyze, plan, and guide their projects. Successful measurement requires a solid understanding of the products, processes, and resources to be measured, an understanding which can only be gained via explicit models. In the MVP Project we are integrating the G/Q/M measurement paradigm with the MVP-L process modeling language in order to guide teams of software developers. This integrated approach is supported by a prototype system, MVP-S, a process-sensitive software engineering environment which offers advanced project guidance using role definitions and measurement data. We motivate the need for measurement, sketch an integration of the measurement and modeling approaches, and demonstrate how MVP-S improves the quality of guidance using measurement.

1 Introduction

A widely recognized problem in software development and maintenance is that organizations do not have intellectual control over their software projects. The lack of intellectual control is visible in an organization's difficulties with planning accurately, with meeting budgets and schedules, and with delivering products that meet the required quality factors. We believe that causes of poor intellectual control over projects include an organization's uncertain understanding of its weaknesses and strengths, continual changes in the marketplace leading to volatile requirements, and inadequate communication among personnel during projects [8].

One approach towards gaining a solid understanding of the organization, managing constant change within a project, and improving communication among personnel is planning and guiding a project using measurement and explicit process models. This approach has found acceptance in the community, as seen in the special issue of IEEE *Software* "Measurement-based Process Improvement" (July 1994). We claim that integrating measurement and modeling will yield synergy effects from which both the group

---

 * Supported by the Software Technology Transfer Initiative Kaiserslautern (STTI-KL).
** Supported by ESSI Project Number 10358, "Customized Establishment of Measurement Programs."

developing a product (the project organization) and the group responsible for corporate process improvement (the Experience Factory, [2]) will benefit. First, quantitative criteria are added to process models to define the successful outcome for a process step; the data collected during the project are used to evaluate these criteria and thereby help developers decide on appropriate courses of action. Second, the validity of collected data can be maximized by using automatic support to request data when the information is fresh in people's minds, to perform some validation automatically, and to provide explanations of the data on demand. Third, feedback can be provided early during process steps when corrective action may be taken [15, 40], not after a process step has ended.

We have implemented MVP-S, a process-sensitive software engineering environment (SEE), that supports our integrated approach for enacting project plans using role definitions and measurement data. This system is the first prototype of our ideas for integrating process modeling and measurement [26]. We give a detailed example of enacting a project plan using role definitions and measurement data to illustrate the advantages of our integrated approach. The reader will gain an appreciation of some issues surrounding data collection in the software engineering domain and an understanding of the benefits offered by project guidance based on role definitions and measurement data. The problem of deciding *what* data to collect is outside the scope of the paper; see [4, 3].

We offer the following definitions to avoid misunderstandings caused by overloaded terminology:

1. Empirical data: objective and subjective data gathered from a project's products, processes, and resources, including data collected by querying people and by using tools to measure work products.

2. Empirical model: a relationship between aspects of products, processes, and resources that was identified using empirical data. These models capture organizational knowledge and are often expressed as formulas. Empirical models invariably depend on the organizational context and may be used to predict such attributes as cost or reliability for future projects that are performed in the same context [11].

3. Role: a set of activities which an agent performs. Lonchamp describes the functionality of a role definition in [23] as "permissions to perform the set of activities [...] and obligations to satisfy the corresponding constraints."

4. Project plan: a description of the processes to be performed during a project, the work products that are consumed and produced by those processes, the entry and exit constraints on those processes, the roles which are responsible for performing the processes, the mapping of individuals to roles, and the resources foreseen for accomplishing the work. We specify entry and exit constraints using empirical data.

5. Guidance: "indirect support to actual software developers or managers working with a process-centered software engineering environment through interpretation of an instantiated process model" [23]. The support may consist of indicating which activities need to be performed or of showing collected data values as compared to target values. Guidance as used here does not include proposing solutions to problems encountered by its users; i.e., no AI-style planning.

6. Process-sensitive software engineering environment (SEE): "a computer system that provides some assistance to its users by interpreting explicit guidance-oriented

[...] software process models" [23]. Our understanding extends this definition in that the system also collects data during the project and provides feedback based on the collected data.

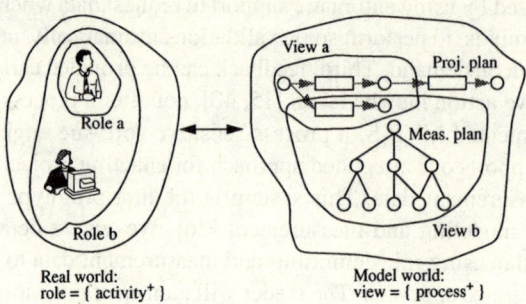

Fig. 1. Interface between real and model worlds

To finish the discussion of terminology, Fig. 1 illustrates our distinction between concepts in the real world (the activities which a software developer performs) and their mapping onto the model world (the abstract representation of the real world). In the real world, a role, as defined above, determines the set of activities which a person may perform. In the model world, a view provides all needed information for a person who performs the processes (activities) associated with the role [39].

Section 2 summarizes SEE research projects which address the issue of measurement in some way. Section 3 expresses our ideas about practical support for data collection in a process-sensitive software engineering environment as a set of requirements. Section 4 presents our prototype system, MVP-S. Section 5 uses a well-known process to give a detailed example of using measurement data to enact a project plan using MVP-S, and Sect. 6 offers some conclusions.

2 Related Work

Many systems support defining and performing processes [9, 33, 19, 13], but only a few support the collection and use of empirical data [24, 25]. A few SEEs which address the issue of measurement in some way, including data collection for understanding and use of data to provide guidance, are summarized here.

The Ginger system [28] consists of a set of monitoring and feedback tools designed to record and improve programmer productivity during coding activities. Data is collected from on-line activities unobtrusively and automatically, and this data is used to provide real-time feedback to the programmers. Measurement support is fixed, and Ginger does not support process definition.

Amadeus [38], more a subcomponent of a SEE rather than a stand-alone system, supports data collection and analysis activities when used in conjunction with a process-enactment system. On-line mechanisms recognize triggers generated by the process-enactment system and respond by invoking tools to collect and analyze data. Amadeus does not directly support process modeling or enactment.

The ES-TAME system [32] is a prototype expert system to support the design process for real-time software. The system's knowledge representation framework supports the development and representation of both work processes and product quality models according to the G/Q/M Paradigm. A particular focus in this system is on using templates to support the construction of measurement plans in the form of a set of goals, questions, and metrics.

The Provence system [22] is a process monitoring and data visualization system that has advanced hooks into the file system of its host computer. Provence offers process definition capabilities via its subcomponent Marvel [18], but the collected data are not used for guiding the process.

The Ceilidh system [5] is a quality control system for teaching students to develop programs according to specifications (functional correctness) and to predefined, empirical quality standards for the code (style, complexity, etc.) Ceilidh supports only the coding activity and collects its data when students submit their code for evaluation. Ceilidh, like Ginger, does not treat the process as a variable.

The SynerVision system [16] supports defining, guiding, and enforcing a process for software development activities by defining and enacting process scripts. Dependencies in the scripts express the order in which activities can be performed; a dependency rule can additionally invoke a tool to check whether a requested action (e.g., marking an activity as complete) is legal. The system automatically measures the time which a person spends enacting a script, but all other data collection and use of data in dependencies is supported by the tools, not the system.

3 Requirements for Data Collection

This section addresses the problem of gathering empirical data from an ongoing software project using a SEE, sometimes called placing "hooks into the real world" [17]. But unlike programmatic hooks, by which some software packages may be customized and which run automatically, data collection hooks often depend on human interaction. Examples of data that must be supplied by humans are the effort spent tracing a system failure to a software fault and the classification of the fault's type and probable cause.

The difficulty of using a SEE to collect data by interacting with people encouraged us to refine a few of the high-level requirements presented in [37] to focus solely on data collection. We state requirements which a process-sensitive software engineering environment must satisfy in order to collect empirical data from personnel during a software project. Our requirements, which were first sketched in [14], are divided into basic technical issues, technical issues due to human involvement, and nontechnical issues. Because this is still an open question that requires further research, we do not claim that meeting these requirements is sufficient for solving all problems in data collection.

3.1 Basic Technical Issues

We understand the process of data collection to be composed of the following steps:

Step 1: Decide upon a data item (metric) and define it unambiguously.
Step 2: Specify the condition that should trigger data collection when it occurs.
Step 3: Recognize when the trigger has occurred.
Step 4: Collect a value for the data item.

Deciding what data to collect and defining the metric unambiguously is a highly creative activity. The task of step 2, specifying a triggering condition for the computer, ranges from trivial to impossible; the recognition of the trigger's occurrence (step 3) is equally easy or difficult. Only people understand trigger specifications such as "detailed design is complete" and can recognize their occurrence. However, it is simple to specify such triggers as saving a file from an editor or compiling a source-code file, and it is equally simple for a machine to recognize their occurrences. For example, in the Provence system [22], a recognizable trigger consists of moving a document into a designated directory. In the case of on-line work, where the work products are available to on-line data collection tools, data collection (step 4) from those products can be automated. This discussion of data collection motivates our first three requirements:

R1, accept machine-understandable specification of triggers.
R2, recognize occurrences of triggers.
R3, invoke tools to collect data.

The SEEs must be coupled with on-line work to justify the preceding discussion of invoking tools to collect data. Fernström introduces four coupling levels in [12], namely loosely coupled (a euphemism for uncoupled; the SEE only knows what people tell it), active support (access to work products is partially automated), process enforcement, (access to work products is totally controlled), and process automation (no human intervention is required). Requirements R1–R3 motivate the next requirement:

R4, the SEE is coupled ("hooked") to on-line work products.

Next we offer a few examples of automated data collection from on-line work products. One involves attaching data collection to a version control system, where the action of checking a document into the system is defined as the trigger. Another example is presented by the tools for conducting inspections of documents on-line [27], which could be used to gather data about the complex inspection process at low cost. Finally, CASE tools for constructing software work products offer a great (yet mostly neglected) possibility for collecting data about those products. For example, the JoYCASE system automatically collects values for the function point and function bang metrics from structured analysis diagrams drawn using that system [36].

Data collection activities can be expected to yield an enormous store of data for which persistence and querying capabilities will be required. Storage, retrieval, and management of empirical data should be delegated to a database management system.[3]

---

[3] This issue should not be confused with the problem of storing all types of software-engineering work products in a database.

Query support systems can then be used to test hypotheses about the data, a function arguably outside the domain of a SEE. This issue motivates the next two requirements:

R5, store empirical data in a database.

R6, retrieve empirical data from a database as needed.

An SEE may obtain data from data-collection tools via two communication paths, either via a direct coupling with a tool or via a database. In the first, the SEE invokes a tool and accepts the new data directly from that tool. In the second, the SEE invokes a data-collection tool that stores the new data in the database. The SEE subsequently uses a unique key (probably obtained from the tool) to query the database for the newly collected data, or may fetch previously collected data from the database without invoking a collection tool. This discussion leads to the next requirement:

R7, accept data from tools via multiple paths.

3.2 Technical Issues Caused by Human Involvement

Data cannot be demanded from a person in the way data can be collected from an on-line product. The person may not be available at the precise moment when the SEE needs the data, or may need a long time to collect and submit the data. Therefore, the SEE must not block or otherwise delay all process guidance and evaluation activities while waiting for a prompt reply from a person. A reasonable solution is that the SEE sends an asynchronous request to the person about the data which it needs. The SEE does not block on the data supplier until the data is provided, but instead accepts the data at any time after sending the request. Because data will arrive at unpredictable times, the SEE must deal with incomplete data on a regular basis. From this discussion we have the following requirements:

R8, notify people of the need for data from them.

R9, accept data from people asynchronously.

R10, function with incomplete data.

In order to collect data about off-line work such as think time, brainstorming sessions, meetings, and inspections, the people involved must be queried. On-line forms offer a reasonable method for querying people without requiring a human interviewer [37]. Figure 2 gives an example of an on-line forms tool, a possible implementation of this requirement:

R11, query people using on-line forms.

3.3 Nontechnical Issues

These issues are primarily concerned with gaining the trust and acceptance of the humans who supply data. Technical personnel must be well informed about the goals defined by management as well as the intended use of the data so that they are willing and able to provide valid data. Providing this information on-line will help people understand the goals and allow them to refresh their memories as needed. These issues, also discussed in [4], motivate the following requirements:

```
┌─                                    prf                                    ·┐
│Press control-g to enter menu. Press control-h for help.                    │
│File  Quit                                                                  │
│                                                                            │
│PERSONNEL RESOURCES FORM                                                    │
│    Name: C. M. Lott                                                        │
│    Project: MVP-S                                       Date: 26 Feb 94    │
│                                                                            │
│Section A: Total Hours spent on Project for the Week:  14                   │
│                                                                            │
│Section B: Hours by Activity                                                │
│           Predesign                                                        │
│           Create design            10▌                                     │
│           Read/Review Design                                               │
│           Write Code                4                                      │
│           Read/Review Code                                                 │
│           Test Code Units                                                  │
│           Debugging                                                        │
│           Integration Test                                                 │
│           Acceptance Test                                                  │
│           Other                                       Go on to Section C y  │
│                                                                            │
│Create Design: Development of the system, subsystem, or components design.  │
│    Includes development of PDL, design diagrams, etc.                      │
└────────────────────────────────────────────────────────────────────────────┘
```

Fig. 2. Forms-based data collection tool (derived from [31])

R12, explain data definitions on-line.
R13, explain goals and intended use of the collected data on-line.

Next we discuss data which is linked to individuals. First, storage of such data is restricted by law in many countries, Germany among them. Second, if management is going to see the numbers and their correspondence with individuals, there is the possibility that future performance reviews will be based on the data, for example on a developer's speed in isolating faults or effectiveness in testing code. In that case, people will do everything in their power to make themselves look good; this is absolutely understandable but it ruins data validity. The ultimate solution is to mask out identities when the data is stored. These issues are discussed at length elsewhere (see [35]) and motivate this requirement:

R14, allow the identities of all data submitters to be masked.

The third and last nontechnical issue is the extra work, intrusion, and annoyance associated with providing data. To address this issue, both management and technical people must be educated about the costs associated with collecting and validating data, as well as the benefits that stem from analyzing the data (for cost-benefit analyses, see [29, 10]). Second, all attempts must be made to minimize the effort of providing data and the intrusiveness on people's work. The computer can reduce the effort of collecting data by automating some collection activities and decreasing the effort required for other activities. The following requirement is also addressed in [22]:

R15, minimize intrusiveness and overhead.

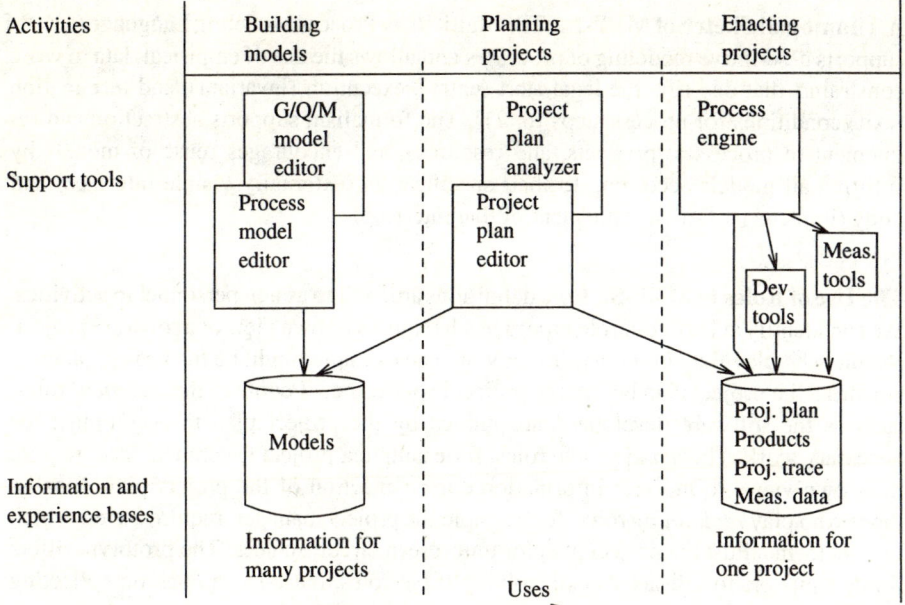

Fig. 3. Scope of activities for a process-sensitive SEE

4 MVP-S, A Prototype SEE

The MVP-S system provides role-specific guidance to software developers during their projects based on explicit project plans, role definitions, quality models, and collected measurement data. Figure 3 illustrates the scope of activities which a process-sensitive SEE might support. Models of processes, products, resources, and measurement activities are constructed using various editors and combined using a project plan editor, thereby forming a project plan which describes the environment's work processes. The project plan may be checked for various properties using a project-plan analyzer, and is finally enacted by a process engine to guide teams of developers. Empirical quality models for the environment's products, processes, and resources are integrated into the project plan to define process constraints and to set goals. During enaction of the project plan, various development and measurement tools are used to accomplish the work and to collect data. All SEE tools must offer a user interface.

4.1 Existing Functionality

In the context of the MVP Project we have developed prototypes for all support tools shown in Fig. 3. The paper focuses on the process engine for MVP-L project plans and the user interface that allows personnel who play technical roles to interact with the process engine.

A Thumbnail Sketch of MVP-L. The Multi-View Process Modeling Language (MVP-L) supports descriptive modeling of processes and allows the use of empirical data to write constraints that describe the legal start (entry), execution (invariant), and termination (exit) conditions for process steps [6, 21]. The formalism supports abstraction and refinement of processes, products, and resources, and encourages reuse of models by splitting all models according to their specification (externally visible interface) and body (implementation or refinement of that interface).

The Use of Roles in MVP-S. Role definitions are used to assign personnel to activities. We can identify at least four role *groups*; each requires its own view of a software project. People who play the *planning* roles, of which an example might be the project planner, construct the project plan before the project is performed. People in the *technical* roles, such as the software developer,[4] are guided by the project plan to accomplish the necessary work. The *management* roles, for example a project's technical lead, require their own views to monitor information during enaction of the project plan. Finally, those who play *replanning* roles, for example the project manager, require a view which will let them adjust the project plan for unforeseen circumstances. The prototype offers limited support for all activities involving these roles, but concentrates on collecting data from and offering guidance to the personnel who play the technical roles.

Integration of Measurement. Measurement and modeling are complementary approaches which enjoy a synergy effect when applied together [26]. We use the G/Q/M Paradigm for defining measurement plans [4, 3], and the formalism MVP-L for writing project plans. The integration of measurement into project plans means defining the goals for all measurement activities that are accomplished in the project, and giving an explicit description of the measurement data which are needed from the project to track any process-improvement goals. Future work will extend the theoretical basis for integrating the two approaches.

User Interface for Technical Roles. People who play technical roles see a view of the project named the role-specific work context.[5] An example of a MVP-S role-specific work context window appears in Fig. 4, which shows the processes for user "lott" who plays the role of a "Design Engineer" in project "ISPW with measurement." Column 1 in the figure shows the names of the activity instances (process instances), column 2 the names of the corresponding activity models (process models), and column 3 shows the status of each activity instance. The status value for an activity instance is one of disabled (can't be performed), enabled (can be performed but no one is doing so), or active (is being performed). Upon receiving a request to start or complete an activity, the process engine checks the entry and exit criteria specified for the process and informs the requestor whether the request was in accordance with the project plan.

---

[4] Includes designers, programmers, testers, quality assurance engineers, maintainers, etc. An alternate term is "software engineer," but we favor "developer" because it is short and specific.

[5] The terms work context and working context are used by various SEEs with dramatically different meanings; see the MERLIN ([34]) and Process Weaver ([12]) projects.

```
 ┌─┐                        MVP-S Developer Interface                              ┌─┐
 └─┘                                                                               └─┘

   action   info   objects   plan   quit


   USER NAME: lott              PROJECT PLAN: ISPW_with_measurement       ROLE: Design_Engineer

   step_major_design_review        Major_Design_Review              disabled
   step_minor_design_review        Minor_Design_Review              disabled
   step_modify_code                Modify_Code                      enabled
   step_modify_design              Modify_Design                    enabled
   step_test_unit                  Test_Unit                        disabled
```

Fig. 4. Role-specific work context window in MVP-S

The task of invoking data-collection tools is delegated to the user interface because the process engine does not necessarily have access to the work products on the user's machine. This is an example of synchronous data collection: in response to a request from the process engine, the interface invokes the data-collection tool and waits for a response. Because the data is directly accepted by the SEE, this mechanism partly implements requirement R7 (accept data via multiple paths). If a forms-based interactive tool is called via this mechanism, this functionality satisfies requirement R11 (interview people using on-line forms). If a tool is used to collect measurement data from artifacts within the computer system, this functionality satisfies requirement R3 (invoke automatic data collection tools).

Process Engine. The MVP-S process engine interprets a textual MVP-L project plan and all related models, creates an internal representation of the plan, processes requests to query and manipulate the current project state, and maintains the project state across shutdowns of the host computer. The process engine waits for events generated by the developers, reacts to the events appropriately, and sends back the result of the events via the information channel provided by the user interface. An event may be a process completion, a change of an attribute value, or a request for information. In the prototype, events are processed serially.

The process engine also meets the requirements of notifying humans of its need for data (R8), accepting data asynchronously (R9), querying individuals using forms (R11), and minimizing the intrusion of data collection (R15). To accomplish all this, we use that ubiquitous system of the information superhighway, electronic mail. When the process engine discovers that it needs data from an individual, it generates a mail message and gives it to the machine's operating system to be delivered. We thereby can reuse both a well-known interface as well as extensive spooling facilities. The individual is expected to respond to the data request and eventually to send it back to the process engine. The process engine receives the mail, parses the message, and extracts the new data.

4.2 Missing Functionality, Possible Extensions

Interface facilities of the MVP-L modeling language are not yet powerful enough to accomplish all of our data-collection goals. Therefore the MVP-S prototype is poorly coupled with on-line work, and automatic data collection from work products is not supported elegantly by the prototype. This means that MVP-S does not completely satisfy requirement R4 (SEE coupled to on-line work products).

We would like to have graphical representations of processes, products, and resources similar to [30] in order to improve comprehension of this information. Given such a representation, the current status of each activity could be indicated graphically, for example by using different colors. The products that are generated by some activity, but do not yet satisfy the exit criteria of that activity, could be highlighted to show the developer which products still require work.

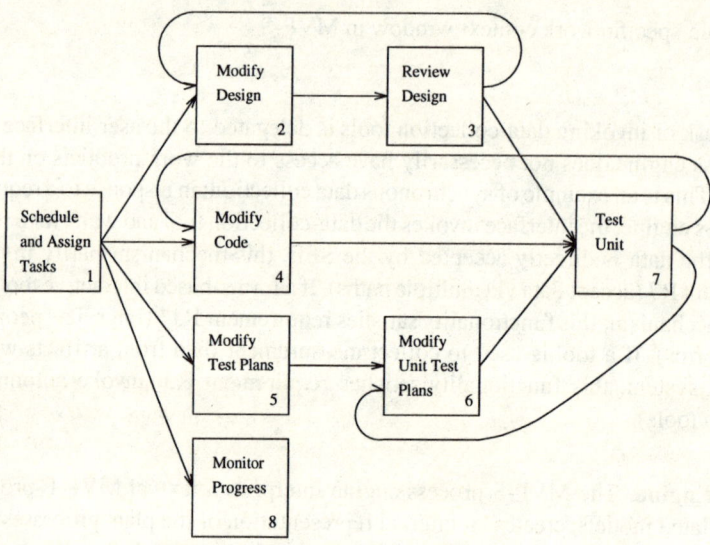

Fig. 5. Control-flow view of the ISPW activities (from [7])

5 Example Project Plan Enaction

We demonstrate the advantages brought by the use of role definitions and measurement by presenting an example of how a developer uses MVP-S to enact a portion of a project plan. For our project we reuse the "Software Process Modeling Example Problem" that was developed for the Sixth International Software Process Workshop [20]. In this scenario, a team of maintainers cooperates to design, code, and test a maintenance change to a software module. Figure 5 presents a view of the project showing the major control flow relationships among the individual process steps. We assume that the organization is interested in learning about faults, their causes, and their repair costs. Empirical models were already developed using historical data to describe characteristics of the software system and their relationship to faults, namely design complexity, expected fault rates in changed code, and effort to implement different types of changes. All of these goals can only be tracked if empirical data is collected from the work activities.

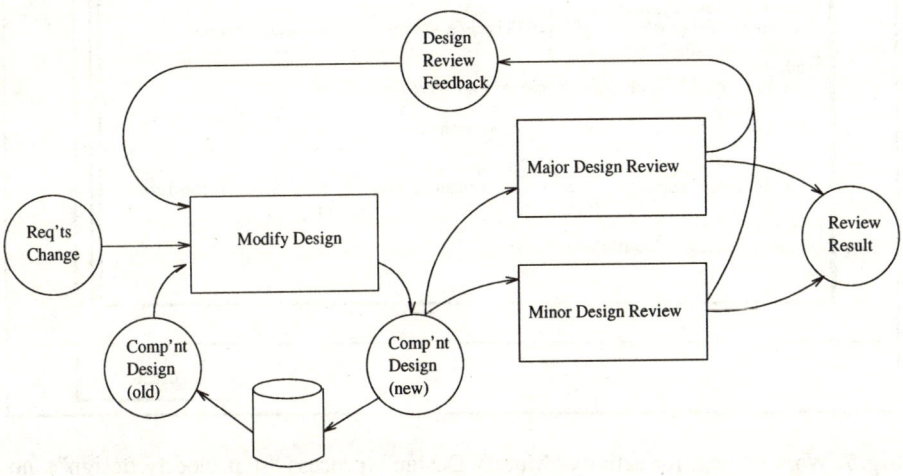

Fig. 6. Data-flow view of the detailed example

We present two variations on this problem that involve enacting the project plan first without measurement data and then with measurement data. Both variations are restricted to the two activities "Modify Design" and "Review Design;" they differ in the information provided for choosing a specific type of review. If the change was small, the review is accomplished without calling a meeting and is named "Minor Design Review." If the changes were large,[6] the reviewers will need to spend extra preparation time before the review and will meet to review the design; this is named "Major Design Review." We refine the original process step "Review Design" into the two steps mentioned above. Figure 6 shows a view of the data flow in the refinement. In the first variation, no support is given for selecting a review process. In the second variation, the design engineers use measurement data to select the most suitable review process of the two.

5.1 Enaction Without Measurement Data

The first variation uses a project plan with simple goals defined for activity "Modify Design" (process "step_modify_design"), namely that its outputs must be provided. Entry constraints for both review processes are also simple, namely that the inputs must be available, so both have the same enabling condition. The criteria which distinguish

---

[6] "Small" and "large" will be quantified in Sect. 5.2.

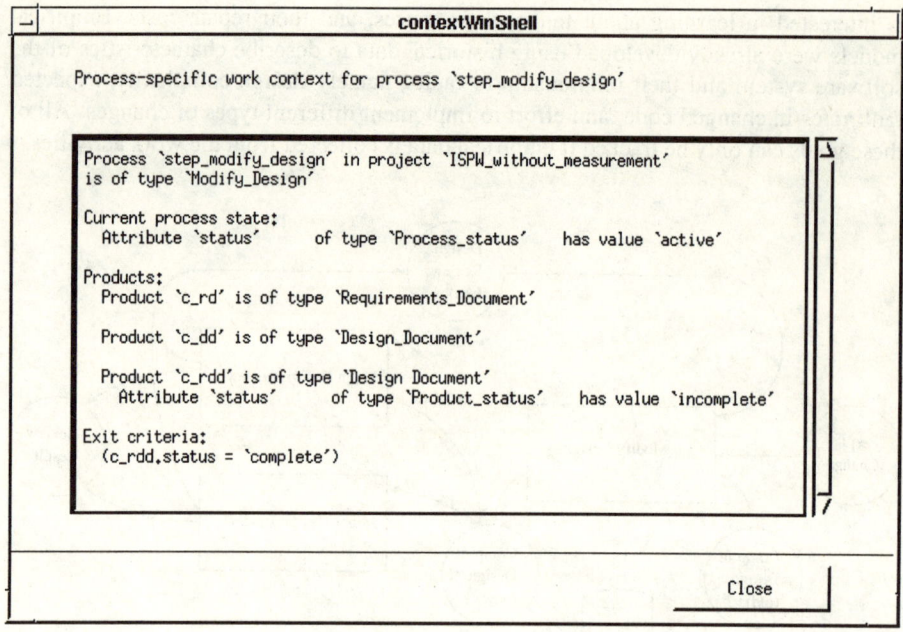

Fig. 7. Work context for activity "Modify Design" (process "step_modify_design"), no quantitative guidance

between small and large changes cannot be represented explicitly without the use of measurement and are therefore missing here.

The project begins with the initial state shown in Fig. 5 after the steps have been assigned. One of the engineers notifies MVP-S that the activity "Modify Design" (process "step_modify_design") has begun by sending the event "start." A design or quality assurance engineer may request additional information from MVP-S about that activity. MVP-S answers with an activity-specific work context showing the products consumed and produced, the current state, and the goal of the activity as shown in Fig. 7.

Eventually the design engineer decides that the activity "Modify Design" (process "step_modify_design") is complete and notifies MVP-S by sending the event "complete" to that process. In the resulting project state, both review activities are marked by the system as enabled because their inputs are available. Figure 8 shows the role-specific work context for a QA engineer, who sees that both design review steps are enabled. The assumption is that the QA engineers will know that only one of the review steps should be selected and enacted. The choice of which step to enact is left to the experience and intuition of the engineers who participate in that process step. We do not wish to argue against individual control, quite the contrary. However, this juncture is a great opportunity for the SEE to supplement the individual's experience and intuition with concrete information about their obligations and which process should be enacted next according to the project plan. This guidance can only be provided to personnel if the organization's implicit knowledge can be made explicit in the project plan.

```
┌──────────────────────────────────────────────────────────────────────┐
│ ─│                    MVP-S Developer Interface                    │ ◢ │
├──────────────────────────────────────────────────────────────────────┤
│  action   info   objects   plan   quit                                 │
├──────────────────────────────────────────────────────────────────────┤
│  USER NAME: agse          PROJECT PLAN: ISPW_without_measurement     ROLE: QA_Engineer │
│ ┌────────────────────────────────────────────────────────────────────┐ │
│ │ step_major_design_review      Major_Design_Review        enabled   │ │
│ │ step_minor_design_review      Minor_Design_Review        enabled   │ │
│ │ step_modify_test_plans        Modify_Test_Plans          enabled   │ │
│ │ step_modify_unit_test_plans   Modify_Unit_Test_Plans     disabled  │ │
│ │ step_test_unit                Test_Unit                  disabled  │ │
│ │                                                                    │ │
│ │                                                                    │ │
│ └────────────────────────────────────────────────────────────────────┘ │
│ ┌────────────────────────────────────────────────────────────────────┐ │
│ │                                                                    │ │
│ └────────────────────────────────────────────────────────────────────┘ │
└──────────────────────────────────────────────────────────────────────┘
```

Fig. 8. Role-specific work context for a QA engineer, no quantitative guidance

5.2 Enaction With Measurement Data

The second variation augments the first with objective, measurable goals for the activity "Modify Design" (process "step_modify_design") and establishes quantitative entry constraints for the two design review steps. Using measurement data to enact this example results in at least three advantages both for the developers and for the organization. First, the state and current situation of the activities can be described more precisely. Second, specific goals of each activity can be presented to the engineers during the time that they perform them. Third, the SEE can use data to provide guidance about which of the two design review steps should be performed after the design-modification activity is completed.

Our choice of constraints is based on the following goals and assumptions. First, data should be collected about design quality aspects to help predict code complexity and maintainability. Second, the organization wants to collect data about the effort (i.e., cost) required to make the change. Third and finally, a review process that is appropriate for detecting defects in that module's design should be chosen; we assume that a minor review is expected to be sufficient when fewer than 30% of the lines were changed, and that a major design must be done otherwise.

Based on these goals and assumptions, we choose to measure the engineer's effort as well as the design quality aspects of module coupling, module cohesion, information hiding, number of lines changed, and the total size; all of these depend on a formal design representation. We specify that effort data must be collected from the person performing the design change upon completion of that activity. We assume that the SEE is coupled to the work environment and can collect data from the design document using measurement tools. Data that can be collected automatically are the size, the number of changed lines, and the values for the coupling, cohesion, and information hiding metrics. Data requested directly from the design engineers is the effort spent in changing the design.

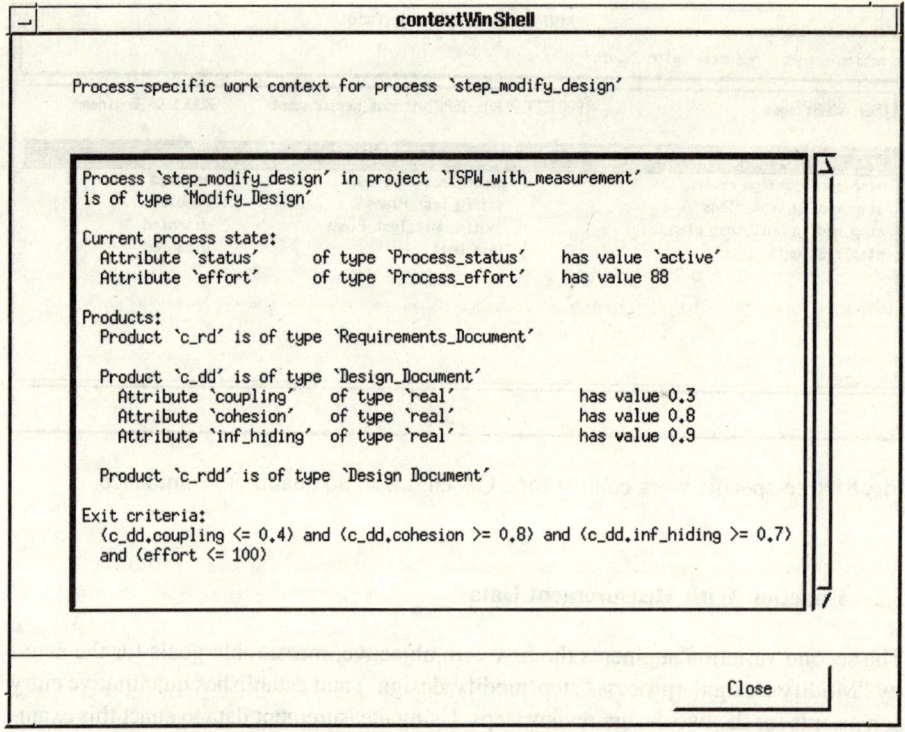

Fig. 9. Work context for activity "Modify Design" (process "step_modify_design"), with quantitative guidance

This variation also starts from the initial project state sketched in Fig. 5. The addition of measurement data makes it possible to provide the design engineer with highly specific information about the design-modification activity after it has been started. Figure 9 shows how the information about the current state and specific goals of the activity "Modify Design" (process "step_modify_design") are presented to a person playing the role of the "designer." The differences with respect to Fig. 7 include showing values for the design-quality attributes "coupling," "cohesion," and "inf_hiding" as well as the definition of the goals of the activity (the exit criteria) using empirical data.

The design engineer eventually judges that the activity "Modify Design" (process "step_modify_design") is complete and notifies MVP-S of the completion of the process step by sending the event "complete" to that process. In response to this trigger, the SEE gathers data about the design product (coupling, etc.) by invoking measurement tools and gathers data about the effort required by sending a mail request. Figure 10 illustrates the mail request for effort data which the process engine sends to the designer.

We assume that data gathered from the design document about coupling, cohesion, and information hiding was within the expected bounds, and that the effort data collected via the e-mail request was also within the preestablished limit. The data for the

```
 ___|                            xterm                            |_|
|Date:      Sun, 14 Aug 94 9:52:48 MET DST
|From:      mvps@informatik.uni-kl.de
|Subject:   Request for data
|To:        lott@informatik.uni-kl.de
|Status: R


                      MVP-S REQUEST
                      -------------

|To: lott
|Project: ISPW_with_measurement
|Role: Design_Engineer
|Authorization: 19940408a3cc4

|Message: amount of effort in hours
|--------------------------------------------------------------------
|Please fill in the value for the requested attribute, restricting your
|changes to the field below, and return this form to the address
|          mvps@informatik.uni-kl.de

|Please enter value for attribute 'effort' of attribute type 'Process_Effort'

|HERE->_____


|Thank you!
|Mail> █
```

Fig. 10. E-mail request generated by the process engine

number of lines changed show that the previously chosen 30% threshold was exceeded. Therefore, the SEE gives the developer guidance saying that the activity "Major Design Review" (process "step_major_design_review") is the appropriate next step according to the project plan. Figure 11 shows the MVP-S window with this feedback. This information can improve coordination among the participants in this process step because all affected personnel understand that, according to the project plan, the more detailed review process is most appropriate for finding defects in the newly modified design document. By using quantitative criteria, implicit information about an activity was made explicit for all personnel.

6   Conclusion

We discussed the integration of measurement and process modeling, gave a number of requirements for collecting empirical data in a process-sensitive SEE, and demonstrated the benefits of using role definitions and measurement data to enact a portion of a project plan. Our prototype system, MVP-S, offers initial support for enacting project plans using role definitions and empirical data, and is the first implementation of our requirements for practical data collection in SEEs. We believe that humans supply extremely interesting and valuable data, and work towards offering automatic support for data collection methods that interact with humans in order to provide measurement-based project guidance. Future work will focus on refining the model for integrating measurement and process models, on developing better models of the feedback necessary for guiding teams of developers, and on supporting organizational learning from the experience gained through the use of our approach.

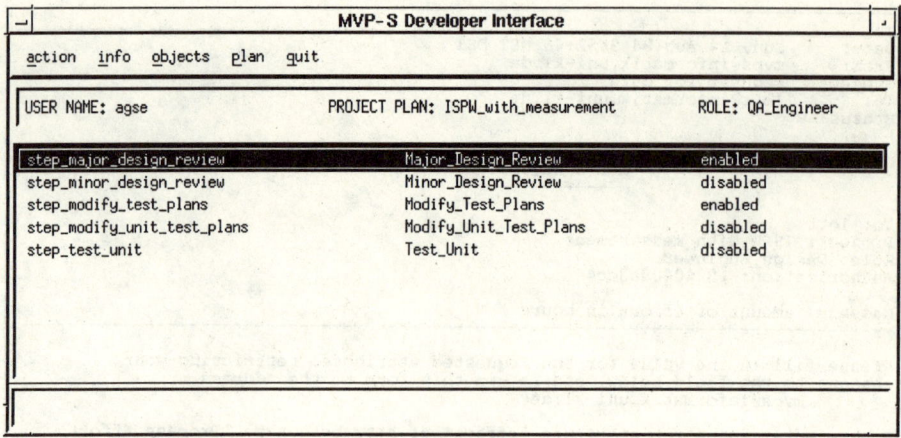

Fig. 11. Role-specific work context for a QA engineer, with quantitative guidance

Acknowledgements

Discussions with Victor Basili, Andreas Birk, Alfred Bröckers, Peter Giese, and Martin Verlage improved the paper significantly. We thank Ralf Kempkens, Enno Tolzmann, and Jürgen Wüst for their work on the process engine and user interface for MVP-S.

References

1. Victor R. Basili. Software development: A paradigm for the future. In *Proceedings of the 13th Annual International Computer Software and Application Conference (COMPSAC)*, pages 471–485, Orlando, Florida, September 1989.
2. Victor R. Basili. The Experience Factory and its relationship to other improvement paradigms. In Ian Sommerville and Manfred Paul, editors, *Proceedings of the 4th European Software Engineering Conference*, pages 68–83. Lecture Notes in Computer Science Nr. 717, Springer-Verlag, 1993.
3. Victor R. Basili and H. Dieter Rombach. The TAME Project: Towards improvement–oriented software environments. *IEEE Transactions on Software Engineering*, SE-14(6):758–773, June 1988.
4. Victor R. Basili and David M. Weiss. A methodology for collecting valid software engineering data. *IEEE Transactions on Software Engineering*, SE-10(6):728–738, November 1984.
5. Steve Benford, Edmund Burke, and Eric Foxley. Learning to construct quality software with the Ceilidh system. *Software Quality Journal*, 2(3):177–197, September 1993.
6. Alfred Bröckers, Christopher M. Lott, H. Dieter Rombach, and Martin Verlage. MVP Language Report. Technical Report 229/92, Fachbereich Informatik, Universität Kaiserslautern, December 1992.
7. R. Conradi, C. C. Malm, E. Lyngra, P. H. Westby, and C. Liu. The EPOS approach to the software process model example problem. In *Collected Solutions from the 6th International Software Process Workshop*, October 1990.

8. Bill Curtis, Herb Krasner, and Neil Iscoe. A field study of the software design process for large systems. *Communications of the ACM*, 31(11):1268–1287, November 1988.

9. Susan A. Dart, Robert J. Ellison, Peter H. Feiler, and A. Nico Habermann. Software development environments. *IEEE Computer*, pages 18–28, November 1987.

10. Raymond Dion. Process improvement and the corporate balance sheet. *IEEE Software*, 10(4):28–35, July 1993.

11. William M. Evanco and Robert Lacovara. A model-based framework for the integration of software metrics. *Journal of Systems and Software*, 26(1):77–86, July 1994.

12. Christer Fernström. Process WEAVER: Adding process support to UNIX. In *Proceedings of the 2nd International Conference on the Software Process*, pages 12–26. IEEE Computer Society Press, February 1993.

13. Alfonso Fuggetta and Carlo Ghezzi. State of the art and open issues in process-centered software engineering environments. *Journal of Systems and Software*, 26(1):53–60, July 1994.

14. P. Giese, B. Hoisl, C. M. Lott, and H. D. Rombach. Data collection in a process-sensitive software engineering environment. In *Proceedings of the 9th International Software Process Workshop*, October 1994.

15. Robert B. Grady. Work-product analysis: the philosopher's stone of software? *IEEE Software*, 7:26–34, March 1990.

16. Hewlet Packard Company. Synervision marketing literature, 1992-1994.

17. Karen E. Huff. Software process measurement session summary. In Wilhelm Schäfer, editor, *Proceedings of the 8th International Software Process Workshop*, pages 18–21. IEEE Computer Society Press, March 1993.

18. Gail Kaiser, N. S. Barghouti, and M. H. Sokolsky. Preliminary experience with process modeling in the MARVEL software development environment kernel. In *Proceedings of the 23rd Annual Hawaii International Conference on System Sciences*, volume II, pages 131–140. IEEE Computer Society Press, January 1990.

19. Anthony S. Karrer and Walt Scacchi. Meta-environments for software production. *International Journal of Software Engineering & Knowledge Engineering*, 3(1):139–162, 1993.

20. Marc I. Kellner, Peter H. Feiler, Anthony Finkelstein, Takuya Katayama, Leon J. Osterweil, Maria H. Penedo, and H. Dieter Rombach. Software process modeling example problem. In Takuya Katayama, editor, *Proceedings of the 6th International Software Process Workshop*, pages 19–29. IEEE Computer Society Press, October 1990.

21. C. D. Klingler, M. Neviaser, A. Marmor-Squires, C. M. Lott, and H. D. Rombach. A case study in process representation using MVP–L. In *Proceedings of the 7th Annual Conference on Computer Assurance (COMPASS 92)*, pages 137–146, June 1992.

22. Balachander Krishnamurthy and Naser S. Barghouti. Provence: a process visualization and enactment environment. In Ian Sommerville and Manfred Paul, editors, *Proceedings of the 4th European Software Engineering Conference*, pages 451–465. Lecture Notes in Computer Science Nr. 717, Springer-Verlag, 1993.

23. Jaques Lonchamp. A structured conceptual and terminological framework for software process engineering. In *Proceedings of the 2nd International Conference on the Software Process*, pages 41–53. IEEE Computer Society Press, February 1993.

24. Christopher M. Lott. Process and measurement support in SEEs. *ACM SIGSOFT Software Engineering Notes*, 18(4):83–93, October 1993.

25. Christopher M. Lott. Measurement support in software engineering environments. *International Journal of Software Engineering & Knowledge Engineering*, 4(3), September 1994.

26. Christopher M. Lott and H. Dieter Rombach. Measurement-based guidance of software projects using explicit project plans. *Information and Software Technology*, 35(6/7):407–419, June/July 1993.

27. Vahid Mashayekhi, Janet M. Drake, Wei-Tek Tsai, and John Riedl. Distributed, collaborative software inspection. *IEEE Software*, 10:66–75, September 1993.

28. Ken-ichi Matsumoto, Shinji Kusumoto, Tohru Kikuno, and Koji Torii. A new framework of measuring software development processes. In *Proceedings of the 1st International Software Metrics Symposium*, pages 108–118. IEEE Computer Society Press, May 1993.

29. Frank E. McGarry and R. Pajerski. Towards understanding software - 15 years in the SEL. In *Proceedings of the 15th Annual Software Engineering Workshop*. NASA Goddard Space Flight Center, Greenbelt MD 20771, November 1990.

30. Peiwei Mi and Walt Scacchi. Process integration in CASE environments. *IEEE Software*, 9:45–53, March 1992.

31. National Aeronautics and Space Administration. Software Engineering Laboratory (SEL) Database Organization and User's Guide, Revision 1. Technical Report SEl-89-101, NASA Goddard Space Flight Center, Greenbelt MD 20771, February 1990.

32. Markku Oivo and Victor R. Basili. Representing software engineering models: The TAME goal oriented approach. *IEEE Transactions on Software Engineering*, 18(10):886–898, October 1992.

33. Dewane E. Perry and Gail E. Kaiser. Models of software development environments. *IEEE Transactions on Software Engineering*, 17(3):283–295, March 1991.

34. Burkhard Peuschel, Wilhelm Schäfer, and Stefan Wolf. A knowledge-based software development environment supporting cooperative work. *International Journal of Software Engineering & Knowledge Engineering*, 2(1):79–106, 1992.

35. Shari Lawrence Pfleeger. Lessons learned in building a corporate metrics program. *IEEE Software*, 10:67–74, May 1993.

36. Raimo Rask, Petteri Laamanen, and Kalle Lyytinen. Simulation and comparison of Albrecht's function point and DeMarco's function bang metrics in a CASE environment. *IEEE Transactions on Software Engineering*, 19(7):661–671, July 1993.

37. H. Dieter Rombach. The role of measurement in ISEEs. In Carlo Ghezzi and John McDermid, editors, *Proceedings of the 2nd European Software Engineering Conference*, pages 65–85. Lecture Notes in Computer Science Nr. 387, Springer-Verlag, September 1989.

38. Richard W. Selby, Adam A. Porter, Doug C. Schmidt, and Jim Berney. Metric-driven analysis and feedback systems for enabling empirically guided software development. In *Proceedings of the 13th International Conference on Software Engineering*, pages 288–298. IEEE Computer Society Press, May 1991.

39. Martin Verlage. Multi–view modeling of software processes. In Brian C. Warboys, editor, *Proceedings of the 3rd European Workshop on Software Process Technology*, Grenoble, France, February 1994. Lecture Notes in Computer Science Nr. 772, Springer–Verlag.

40. Edward F. Weller. Lessons from three years of inspection data. *IEEE Software*, 10:38–45, September 1993.

Combining Process Models and Metrics in Practice

Tineke de Bunje, Alison Saunders

Philips Research Laboratories
Prof. Holstlaan 4, 5656 AA Eindhoven, The Netherlands
e-mail: bunje@prl.philips.nl saunders@prl.philips.nl

1 Motivation and relevance

It is widely accepted that continuous improvement of the software process is crucial to maintain the competitive edge. Modelling the process and collecting metrics are both means to improve the effectiveness of the process [6] and [1]. Instead of maintaining a separation between process modelling and metrication [1], a co-operation between the two activities is proposed similar to [3]. Practical experiences are presented to support this proposal. These experiences have been gained in our own research environment and in an industry group working on TV development.

2 Combining process modelling and metrics

Metrication is often based on the Goal-Question-Metrics (G-Q-M) paradigm discussed for example in [5]. The initial goals are usually established at the highest level of the organisation. These goals are then broken down in sub-goals, sub-sub-goals etc. Questions are related to lower level goals and then both process and product metrics defined to answer the questions. However, this can easily result in a goals tree with too many measurements to be of practical use.

Process modelling relies on the Entry-Task-Exit (ETX) paradigm as explained in [2] and [4], in which processes are described by means of boxes, representing activities. These activities are broken down in sub-activities, sub-sub-activities etc. This can also result in a tree which structure is no longer clear since there are too many redundant details involved.

In practice it is necessary to identify the most important branches in the goals tree and in the activities tree. This can be achieved by combining metrication and process modelling as shown in figure 1. In addition, integrated metrics and models will contribute to a better process improvement programme.

Metrics and process models mutually profit from each other when the metrics support the models and vice versa. Mistakes are often made when choosing and interpreting metrics without a formal process model, while it is very difficult to validate the model of the current process without metrics. Moreover, the success of corrective actions based solely on the collected metrics is likely to be more random than if they also

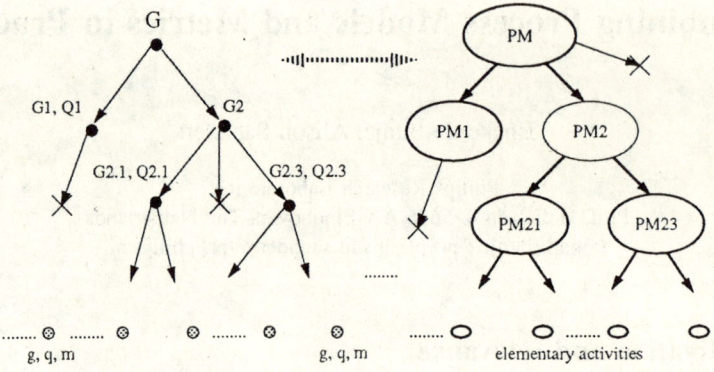

Fig. 1. Interaction between metrication and modelling

refer to a formal model. Similarly, predicting the results of certain changes is likely to lack control if they are solely based on a process model.

Summarizing, a couple of a process model and metrics in which the metrics support the model and vice versa is more valuable than an isolated model or isolated metrics.

Besides the benefit obtained when models and metrics correspond to each other, the activities of modelling and metrication may profit from each other too. The development of the process model and the development of a metrication plan are both characterized by iterative refinements. The refinements can be interleaved and the decisions taken can be inspired by each other.

Stating goals and asking questions, according to the G-Q-M paradigm focusses the process modelling on those subprocesses that attribute most to the recognized goals. The refinement of goals and questions can be guided by the precise formulation demanded by the process modelling. Also, a premature process model may help when identifying attributes needed for metrication. The interaction between process modelling and metrication is illustrated by the practical experiences below.

3 Practical experiences

As a direct result of software process improvement activities, our research group started enthousiastically collecting all sorts of metrics but soon discovered that unless this was approached systematically, a lot of effort was invested without any visible effect. Therefore, after several iterations, the group produced their first metrication plan based on the Goal-Question-Metrics paradigm. This plan was phased, documenting both metrics that were readily available and metrics that were desirable but were currently impractical to collect.

Regular meetings have been introduced to update the plan as more experience is gained and to interpret the measurements that have been collected over the previous months. In this way all group members are able to see that the measurements are indeed being used. During these meetings it was discovered just how important a process model is. Initially a number of metrics had to be scraped because they were measuring the wrong things, i.e. the software process was not fully understood. The converse was also true. By collecting metrics, differences between the model and reality were highlighted.

Using the experiences gained internally, a combined approach of modelling and metrics is currently being applied to the inspection process of a group set up for TV development. In this case building a process model is motivated by the need for decision support based on a structuring of the metrics that are already collected. The formalism in which the ETX-boxes are described is chosen such that it corresponds best to the experience in the industry group.

Up to now, seven main steps have been recognized:

1. Formulate the main goals and distinguish between the product aspects and process aspects of these goals.
2. Make a description of the process as one ETX box, including roles, constraints, conditions and products.
3. Establish the main goals in terms of the black box description of the process and associate questions and possibly also metrics with these goals.
4. Refine the process in about 5–8 subprocesses, guided by the established goals and describe these subprocesses as ETX boxes.
5. Refine the goals and establish questions and possibly metrics for these subprocesses; investigate if behaviour of the whole process can be derived from the behaviour of each of its parts.
6. Refine those subprocesses further that contributes most to the questions.
7. Associate metrics with all relevant boxes and express the answers of the questions in terms of these attributes.

These steps demonstrate clearly the mutual influence of process modelling and metrics: each step from G-Q-M is followed or preceded by a step from process modelling.

As a first step, the following goals were formulated:

– Goal 1 : Improve the product quality, i.e. minimize defects and maximize the satisfaction of the users of the inspected documents;
– Goal 2 : Improve the process quality, i.e. minimize effort, optimize schedule, maximize satisfaction of the process owner.

The second step leads to a black-box description of the process given in figure 2.

Here the following abbreviations are used:

D   : the document to be inspected
HOD   : higher order document(s)
IP   : the inspection process

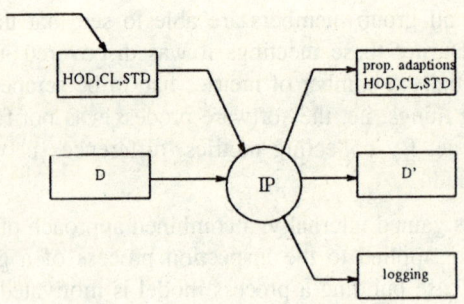

Fig. 2. Black box model of the inspection process

CL   : check list(s)
STD   : standard(s)
D'   : the improved document

Notice that this model is not yet complete, since no roles and constraints are given. In practice, the results of the first few steps are refined once more experience has been gained during the following steps.

The third step was combined with a transition to questions instead of goals. The following questions were associated with the process model and the goals resulting from step 1 and step 2.

- How can the quality of D' be improved?
- How can the quality of the logging be improved?
- How can the quality of IP be improved?

Examples of metrics that were associated with these questions:

- the number of defects of a certain kind (major, minor) found in D;
- the time spent on discussions about logging data during rework;
- the quality of D' compared with the quality of D.

Subsequently, the inspection process was refined in the following activities: prepare the inspection, plan, preparation by inspectors and moderator, logging, rework and closing. For each activity the different roles were given, as well as the input and the output documents (step 4). The goals and questions were refined for each of these activities and the overall goals and questions were related to the refined goals and questions (step 5). Furthermore, the metrics that were already collected were investigated and, when possible, associated with the model. The connection between the metrics and the model is not completely established (step 6 and 7), so the experiment is not finished yet.

4 Conclusions

Practical experiences carried out so far, show that it is plausible to combine metrication and process modelling and thereby profit from their mutual influence. Tooling has not played an important part in the experiences. However, it was found that a simple drawing tool and a spread sheet are useful.

These experiences are being investigated further to validate the approach. It is too soon to comment on the benefits and costs involved. At least part of the benefits is a better understanding of the processes that are involved in software development and a means to measure improvement.

References

1. V. Ambriola, R. Di Meglio, V. Gervasi, B. Mercurio. *Applying a Metric Framework to the Software Process: an Experiment*. In *Proceedings EWSPT'94*, LNCS 772, pages 207–226, 1994.
2. W.S. Humphrey. *Managing the Software Process*, Addison-Wesley, 1989, ISBN no: 0-201-18095-2
3. C.M. Lott and H.D. Rombach. *Measurement-based guidance of software projects using explicit project plans*, Information and Software Technology, vol. 35, no. 6/7, pages 407–419, 1993.
4. R.A. Radice, N.K. Roth, A.C. O'Hare, Jr. and W.A. Ciarfilla *A programming process architecture*, IBM Systems Journal, vol. 24, no. 2, 1985.
5. John Roche and Mike Jackson *Software measurement methods: recipes for success?* Information and Software Technology 36(3), pages 173–189, 1994.
6. Tom Rodden, Val King, John Hughes, Ian Sommerville. *Process Modelling and Development Practice*. In *Proceedings EWSPT'94*, LNCS 772, pages 59–64, 1994.

Application Experiments

Nazim H. Madhavji

School of Computer Science, McGill University & CRIM

tel: (514) 398-3740

madhavji@opus.cs.mcgill.ca

There are a number of specific thrusts in the software process field today, both in academia and industry. For example, we have witnessed a number of efforts aimed at: building process-centred environments, designing process modelling and programming languages, and defining terms and concepts, building alternative capability maturity models, designing process assessment methods, and carrying out process improvements.

An important realisation that is emerging from all this experience is the need for empirical work in the field. In particular, some in the process community are beginning to question the validity of a research solution (e.g., a new method or tool), a finding (e.g., a phenomenon) or a practical result (e.g., process improvement), especially when these have not resulted from the use of rigorous and scientific work methods.

This is an important concern because a considerable amount is at stake once we accept a research or practical result. For example, we start using research solutions in the hope that our practical problems shall be solved; we make important business or other decisions based on findings; and we start trusting that practical results will lead to improved software quality, higher productivity and reduced development costs.

In this session, then, we shall focus on empirical methods as applied to software processes. In particular, we shall consider experiments, case and field studies, and surveys.

The specific goals of this session are: (a) to recognise the growing importance of empirical work in the process field; (b) to identify key issues and approaches in this type of work; (c) to draw some lessons learnt from past experience; and (d) to share knowledge gained thus far. Achieving these goals would be a significant milestone in the history of this workshop.

Beginning with an introduction to this session, we intend to have a presentation based on past empirical work. This shall be followed by an open discussion focusing on the stated goals. We shall conclude with a summary of the key points from the session.

Space Shuttle Onboard Software (OBS) Development and Maintenance Process Automation

Authors: L. B. Strader, M. K. Aune, J. A. Rodgers, M. A. Beims

OBS History and Mission

The Space Shuttle Onboard Software project has been in existence for nearly 20 years. In 1989 the project was rated at the highest level of the Software Engineering Institute's (Carnegie Mellon University, USA) Capability Maturity Model. The high quality software produced by the project is directly linked to its maturity. In addition to the highest level on the Software Process Capability Maturity Model, this project has received several quality and productivity awards. This organization was the first contractor to receive the prestigious NASA Excellence Award and the only project to receive this award twice. The Houston site was twice named the IBM Best Software Lab and was twice awarded the Silver Level in an IBM internal assessment matched against the United States of America's Malcolm Baldridge National Quality Award criteria.[1] In December 1994, the project received unconditional ISO 9001 certification.

This project's commitment to eliminate all software errors led to an incremental, cumulative development of processes that effectively took Problem Reports through a reproducible cycle to Problem Resolution.

The processes to develop the Space Shuttle software evolved over many years. Maturity grew out of practical experience and innovative ideas from industry and academia, as well as through trial and error. Disciplined application of the resulting processes produced software systems that have been virtually error free. Strong program management, adherence to the process even during times of pressure, and procedural disciplines have had a significant positive influence on project results. *None of these factors required sophisticated technology.* They are organizational, procedural and cultural, and can be implemented by managers and software professionals who have the desire to improve their development environment.[2]

The question is, with such a high degree of quality and cost saving measures in place, why automate this process now?

---

[1]C. Billings, J. Clifton, B. Kolkhorst, E. Lee, and W. B. Wingert, "Journey to a Mature Software Process", IBM Systems Journal, 46-61, (Vol. 33, No. 1, 1994)
[2]Billings et. al.

Process Automation

Motivations
The prime motivation in any business activity is to maximize profit. Methods of maximizing profit include reducing costs in an existing process, or by creating a new opportunity. The Onboard Shuttle is an existing process, so the prime motivation at this point is reducing costs using process automation.

Specifically, this project desires to: eliminate excessive paper use, eliminate duplicate tasks, streamline and simplify processes, ensure consistent process implementation across the organization, and simplify maintenance and revision of the processes.

Disadvantages
It is possible to inadvertently increase costs by automating. Some caveats for automation include: tools must preserve the current process, complex tools require more training, expensive tools require capital investment, development resources for home grown tools must be available, and the process must be significant enough to warrant automation.

Deciding Factors
The factors that overcome these possible disadvantages for certain technologies are reduction in purchase price due to exponential reductions in LAN and Personal Computer costs, reduced training costs due to new personnel with school experience and exposure of existing personnel to new technologies in day-to-day tasks (e.g., interoffice memos), and improved perceived cost effectiveness. This perception of cost effectiveness for new technical skills changed for individuals because of the changing work environment outside this project, and for the project because the current contract rewards cost reductions that preserve high quality.

Overcoming Preconceived Ideas
The experience of the OBS project is, that process automation, or enactment, does not necessarily require re-engineering the process. It does not require complex or expensive tools, and can be implemented with relatively simple Commercial-Off-The-Shelf (COTS) applications. Finally, process automation can be done rapidly through prototyping and spiral development.

Process Automation Tool Development

We have found that processes deserving automation exhibit one or more undesirable characteristics. These characteristics include inconsistencies in the format of the finished product, individual use of excessive resources to create the product, products that require maintenance of the mechanism used to generate the product, no sharing of data, no reuse of previous products, and repetitive manual mechanisms such as manipulation of data. In order to add value to a program without interrupting the flow and quality of work, developing process automation tools should abide by the following set of flexible rules.

Leave a working process intact.

The most important rule for process automation to this project is that the process tool needs to fit into or mimic the existing process. To do this, first the process needs to be known, understood, and defined. If the process is well documented, as in an ISO 9000 certified process, the requirements for the tool may be lifted straight from existing documentation. The tool is not to change the process, but simply to automate it efficiently and effectively. If the tool mimics the process, then there will be no shutdown of the current operations for users to understand the flow and logic behind the tool.

Select appropriate tasks for automation.

After the process is well understood, the tasks that are to be automated require definition. The tasks this project considered for process automation are: repetitive tasks, notification tasks, delegation of assignments, tracking of status, and the scheduling of assignments and/or due dates. These tasks need to be either already in place and well documented, or determined to exist through user interviews. The elimination of repetitive tasks through the use of the process tool will have a significant benefit to the users and the organization. Therefore, identifying repetitious tasks through process documentation review and user interviews is essential. This project experienced good results identifying and prioritizing requirements using computerized brainstorming. If the repetitive tasks are automated first, this will give the users a noticeable change in their workload and create an atmosphere of support for use of the current tool, updates to the tool, and future tool creation. This also allows the users to expand the scope of their work as more of the menial tasks are eliminated from their day to day activities.

Additional tasks to consider for automation are those tasks that notify the users when to perform an action, when to record the action, when to distribute an action, and when to close an action. Notification may be currently performed in a number of different fashions from paper actions distributed in mail folders, phone calls, electronic mail, word of mouth, etc. If the process tool incorporates this notification of actions and the status of these actions, then this notification will be global across the process and conform to a set of standards. This allows for a central distribution point without losing any notifications. Scheduling of tasks is another part of any process that needs to be considered for automation. The aspect of when a task is assigned and due is in essence part of the notification process.

Make the new tool very easy to use.

The tool's platform must be accessible to at least as many users as the current process. The platform must be as easy, or easier to access than the current process. Training must also be minimal for the selected platform. On the OBS project, experience shows that a new tool will not be willingly used if it is as difficult, or more difficult to use as the current process. Fortunately, if the tool does mimic the current process, users inherently understand the tool.

Keep the tool development team small and move fast.
Once the process flow and requirements are generated and approved, design and development may begin. It is essential that the design and development teams are kept small relative to the size of the process that will be automated. A team small in size will be able to quickly move through the development process and rapidly prototype the tool. A prototype that can be rapidly delivered to the requirements owners and the users creates an open line of communication. This protects the tool from becoming a cutting edge technology demonstration without practical use.

The user must control the tool.
After the tool is placed into production, the users must be made owners. The users must have an imbedded process of feedback to the tool maintainers. If this is in place, and the configuration management is done by the process owner, then fast implementation of concepts and changes requested by the users can be made. Users are more willing to suggest changes if change requests can be made through the tool and they have direct feedback from the developers. Again, if the users feel that they have ownership in the tool, they will accept modifications and new tools readily, allowing for future process improvement.

Stay with a successful approach for your process.
In order to continue to produce a consistently high quality product, the above set of rules and forms should be followed in designing and implementing process automation tools for existing operations.

Lessons Learned

Requirements definition can be a lengthy and complex task. To achieve the best results, requirements should be formulated by a small group of knowledgeable users and finalized in a reasonable period of time. Tool capabilities and functions should be defined and reviewed incrementally. Attempts to define the whole system and obtain the consensus of the entire user community at once are typically not successful. This leads to too many requirements being defined by too many people at the same time. With incremental requirements definition, the user community can focus on one aspect of the proposed tool and finalize that portion of the tool before moving on to the next function. The best approach is to define and baseline the core functions, followed by the user interfaces, inputs and outputs, and then finally the maintenance and configuration requirements.

Tool development also benefits from an incremental approach. Attempts at developing and deploying large, complex tools may require a very large group of people, excessive amounts of time and resources, and typically result in initial user dissatisfaction. An approach of developing core functionality with incremental enhancements provides for rapid development and implementation, user evaluation, improved maintenance, and easier documentation.

Once a tool has been defined a prototype development phase should follow. A prototype will allow for experimentation, user feedback, and confirmation of requirements. This prototype phase can also be performed outside the Configuration Management practices employed on fully developed tools. Once a prototype has evolved into a stable and meaningful state, has had a chance for user testing and feedback, and has been documented into a refined set of requirements, then place it under configuration control and review. There is ample time to document, control, and baseline a tool once it is ready for production.

In summary, process enactment can be accomplished rapidly, cheaply, effectively, and successfully through the use of small knowledgeable and skilled teams using incrementally defined requirements and prototype development. Tools can be built through creative use of COTS applications, without great complexity or the need to re-engineer processes to fit the tool. Whatever approach is taken, process enactment must be weighed against the need, cost, and benefit of automating a particular process.

PM Case Studies: A Tentative Characterisation

Ian Robertson
email: ir@cs.man.ac.uk

Department of Computer Science
University of Manchester M13 9PL

Abstract. Case studies are an essential feature of process modelling research. This work introduces a possible structure to assist in assembling and reporting the fragmented findings of such experiences

1 Introduction

The Informatics Process Group has been involved in various aspects of process modelling for a number of years. One ongoing thread of this work is our involvement with operational research activities in commercial, industrial, and government organisations. These studies have been undertaken for a number of reasons, primarily as a way of testing our current version of our Process Analysis and Design Methodology (PADM) [?], but also to seek new insights into the pattern of activity which we refer to as 'process'.

One of the features of these studies is the fact that they are very much driven by the client and the learning arising is thus in many ways unsatisfactory: fragmented, sometimes seemingly not very relevant, inconclusive, and frequently incomplete. It is thus difficult to assemble the results of the case studies into a meaningful 'whole' of experience.

These findings thus need to be classified in some way, and this characterisation is only a first step in achieving this aim. Earlier work by Lonchamps [?] and Humphrey and Feiler [?], focussing on the software process domain and the subject process itself, provide a useful starting point to from which to develop such a structure.

In this work, 'characteristics', 'properties' and 'features' are taken to be synonymous. This position paper proposes a characterisation framework for IPG case study experiences, with the Requirements identified in Section 2, and the framework itself outlined in Section 3.

2 Requirements

The ultimate objective is to find a way to organise our knowledge of this area of process experience such that it can be useful and readily accessed by workers. We want:
- To provide a concise yet insightful description of a study.
- To be better able to synthesise our knowledge.

– To identify the essentials in the complexity.
– To be able to recognise a few salient features of a process, and from these few features predict the existence of others.
– To identify features which have a critical effect on both process development and process support.
– To explore the factors which lead to difference between processes.
– To have groupings for easy reference.

As a first step in achieving these objectives, we need a systematic way of identifying features of interest.

3 Study Features

These are the features which help characterise the activity of, and the conclusions that can be drawn from, a case study. They have been selected to represent concerns identified both by earlier workers and by IPG field work. Our work so far has not been able to assess their relative significance, so, at present, their ordering does not reflect importance. They are also chosen so as to be as 'orthogonal' as possible. The framework is illustrated in Figure 1 and the essentials are described below.

3.1 Learning

The enhancement of our knowledge and experience arising from case study field work can be achieved in one or more of four areas: Method, Tools, Enactment Technologies, and Organisations. 'Method' does of course primarily refer to the testing and improving of PADM itself, but the other areas are more general: identifying requirements for new tools, experiences with using existing tools; experience with using particular enactment technologies; determining their appropriate use; the mapping of technologies to particular process characterisations or 'types', and lastly understanding how organisations shape a process, and the extent to which this needs be the concern of the process engineer.

3.2 Purpose

There are usually a number of possibly conflicting reasons for carrying out a study. They do, however, fall into relatively few categories. Different parties to the study do, of course, have their own agendas which can be in conflict, and the existence of differences should be appreciated. An understanding of study purpose can help with the understanding of outcomes. A study may be done simply to describe the process, either 'as is', or 'as ought to be'. Often, a process is to be evaluated according to some established criteria, to determine if the possibility exists for significant 'performance' or 'quality' improvements. If a process is to be changed, it is often acknowledged that a study can promote greater understanding of 'what is', and 'what ought to be', and in fact can assist with the change process itself.

A common IPG purpose is to gain experience in use of the PADM Methodology, and to pass on a knowledge of process engineering techniques to practitioners, thus placing them in a better position to address process-related issues.

3.3 Process Domain

This sets the human, machine and historical context in which the study was carried out, the essential frame of reference for a viewpoint from which to assess the work of the study. The real world can be exceedingly complex and thus it is unfortunately necessary to generalise with a broad-brush description. A prerequisite is to have an understanding of the education, background and career structure of participants, and an appreciation of the unwritten 'rules' of the organisation which can indicate hurdles that will have to be overcome before certain kinds of change can be tolerated. It also seems important to know the extent to which process practitioners understand what they do and why they have to do it, and the existing systems and tools that are used in the process.

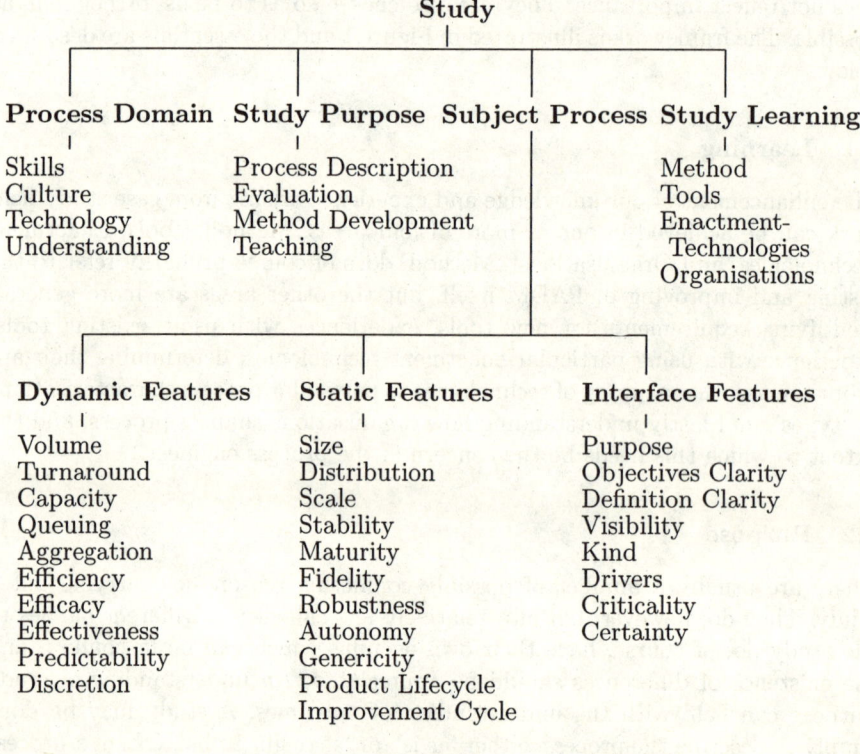

Study

Process Domain   **Study Purpose**   **Subject Process**   **Study Learning**

Process Domain	Study Purpose		Study Learning
Skills	Process Description		Method
Culture	Evaluation		Tools
Technology	Method Development		Enactment-
Understanding	Teaching		Technologies
			Organisations

Dynamic Features   **Static Features**   **Interface Features**

Dynamic Features	Static Features	Interface Features
Volume	Size	Purpose
Turnaround	Distribution	Objectives Clarity
Capacity	Scale	Definition Clarity
Queuing	Stability	Visibility
Aggregation	Maturity	Kind
Efficiency	Fidelity	Drivers
Efficacy	Robustness	Criticality
Effectiveness	Autonomy	Certainty
Predictability	Genericity	
Discretion	Product Lifecycle	
	Improvement Cycle	

Figure 1

3.4 Subject Process

Most of the static and dynamic properties relating to the subject process are those already identified [?] [?]. In some cases it was felt that the definitions were rather too pre-disposed towards automated process execution, and it is hoped to re-examine these definitions in the near future. In the meantime the existing definitions can be used. The most important new property introduced is that of 'Genericity': a measure of the potential for process re-use.

It was felt necessary to introduce a new group of so-called 'interface properties'. These are features which are manifest at the interface of the process model and the performing process, and particularly concern the way the process as defined relates to human participants. These component properties are as follows:

- Purpose: The ostensible purpose of the process may well seem clear, however the situation frequently arises that there are many underlying objectives which ought to be brought to light. The objectives should not be presumed as being 'static', and their relative importance can change, possibly on a day-to-day basis.
- Clarity of objectives: The ease with which different objectives of the process can be distinguished
- Clarity of definition: The ease of understanding of any process definition,
- Visibility: The awareness of participants as to what the process is.
- Kind: Identification by purpose, e.g. an approval process, a design process.
- Drivers: The source of those parameters of the process by which success is measured.
- Criticality: A measure of how important its is to the organisation that the process is performed correctly according to the 'definitions'.
- Certainty: A measure of the likelihood of unforeseen exceptions having to be handled.

4 Conclusions

One of the key features by which a specialism is identified as a discipline in its own right is the degree to which its vocabulary is shared by practitioners. Workers in the field of process modelling are only now beginning to identify and seriously debate the vocabulary which we use. The cited works can thus be recognised as milestones. These were only partially useful when attempting to describe our experiences of case studies. They do, however provide a very sound basis on which to build this tentative structure which we hope will prove useful to process modellers. It is offered for criticism of features which are needed but omitted, and for those which are less important and unworthy of inclusion.

Many of the features described are intuitive and it may well prove difficult to ascribe them meaningful values. It is hoped that this can be explored in the near future as the studies already undertaken by the IPG are cast in this framework. This may well lead to some properties being re-defined, some dropped altogether, or new ones introduced.

This work serves only to introduce the issue of the characterisation of the processes themselves, a small step towards fully addressing the Requirements outlined in Section 2. This is a potentially valuable area of inquiry in that, in all the complexity of organisation processes, there may indeed be only a relatively few distinct processes. Research activity can thus focussed on these few processes with beneficial results.

Work has already been done on what may well be a closely associated area: that of the characterisation of technologies [?] [?]. It seems very unlikely that all processes can be effectively supported by all technologies and some kind of mapping of process properties to technology properties would be of immeasurable benefit to practitioners.

The SERC funded project 'Process Engineering Framework' is currently exploring these, and other areas. This work will be further developed at Manchester by using this framework as a comparative framework incorporating those earlier case studies which have, to a large extent, gone unreported, and also to support and augment more general operational research studies [?].

References

1. A.M. Christie. A Practical Guide to the Technology and Adoption of Software Process Automation. Technical Report CMU/SEI-94-TR-007, Software Engineering Institute, March 1994.
2. P.H. Feiler and W.S. Humphrey. Software Process Development And Enactment: Concepts And Definitions. In *Proceedings of the Second International Conference on the Software Process*. IEEE Computer Society Press, 1993.
3. J. Lonchamp. A Structured Conceptual and Terminological Framework for Software Process Engineering. In *Proceedings of the Second International Conference on Software Process*. IEEE Computer Society Press, 1993.
4. J. Lonchamp. An Assessment Exercise. In A. Finklestein, J. Kramer, and B. Nuseibeh, editors, *Advances in Software Process Technology*. Research Studies Press, 1994.
5. D.G. Wastell, P. White, and P. Kawalek. A methodology for business process redesign: experiences and issues. *Journal of Strategic Information Systems*, 3(1):23–40, 1994.
6. P. White. Report on the Process Analysis and Design Methodology. Technical Report 142, IOPT, 1993.

A Survey and Comparison of Some Research Areas Relevant to Software Process Modeling

Terje Totland, PAKT, and Reidar Conradi, NTH
University of Trondheim, Norway

1.   Introduction

Few research areas have received so much interest from the software engineering community during the past decade as the software production process [Curtis 92]. The main objective of this increase in interest is to improve software production in terms of increased product quality, reduced costs and reduced time-to-market.

The Software Engineering community seems to agree that a fruitful way to support the software production process is through the application of *process models*. Extensive literature on the issue is available, and for a comprehensive "state-of-the-art", see [Promoter 94]. However, the SE community is not the only one focusing on process technology as a means to improve work. This paper presents a brief overview of other communities doing software process technology related research, and points out some main similarities and differences between the research areas.

2. Dimensions for Comparison of Research Areas

There are at least three dimensions that can be used for comparison of the various research areas: Business domain, intent and process elements. By *business domain* is meant the main business area of the organizations that are modeled. The business domain greatly influences the requirements to process technology in each area. Business areas include software production, manufacturing, insurance etc. By *intent* is meant the primary objectives of applying process technology to a specific business domain. The purpose of process modeling ranges from individual understanding of the process to full enactment. Last, process modeling within disparate business domains and with different intent requires focus on different aspects (*process elements*), such as objectives, activities, artifacts, roles, actors, tools etc.

3.   Some Research Areas Utilizing Process Technology

Presentation of each relevant research area is very brief due to the nature of this paper.

3.1   Software Engineering (SE)

Software Engineering is "the establishment and use of sound engineering principles in order to obtain economically software that is reliable and works efficiently on real machines" [Naur 69]. The field of Software Process Modeling can be considered a subdomain of Software Engineering.

Obviously, the business domain of SE is software production. The intent of applying process technology ranges from understanding and visualization of the process to partly automated enactment. Process elements include activities, artifacts, roles, and tools. Artifacts are usually not modeled in depth, i.e., there is no data modeling in the traditional sense.

Both the actual production process and the *metaprocess* is modeled. By metaprocess is meant the building and maintenance of a process model and its related activities. Handling the metaprocess is important in SE, as the production process changes frequently and the process models should evolve likewise. To date, emphasis on the importance of the metaprocess is particular to SE.

3.2   Information Systems Engineering (ISE)

Information Systems Engineering can be defined as application of a set of systematic engineering approaches to develop information systems. An *information system* is a system of computer components, software components, and human and organizational components that are developed, trained and assembled to fulfill the information processing requirements of a problem [Sølvberg 93].

The business domain is usually information services, like banking and insurance. The intent is understanding, analysis and communication of domain knowledge in order to construct the information system. Monitoring, measuring and enactment of the process is not in focus. Process elements are as for SE, except that ISE traditionally has incorporated more of business rules and human actors into their models than SE. ISE also models the artefacts in more detail (data modeling).

3.3   Enterprise Modeling (EM)

Enterprise Modeling is a term that has emerged during the past few years. The term has been given various definitions, but one that seems to be representative for most applications says that "Enterprise Modeling is the process of understanding a complex social organization by constructing models" [Rumba 93]. Hence, the main focus seems to be on *modeling* as a process, and not on the *model* as the important outcome.

EM is *domain free*, meaning that it is not constrained to any particular business domain. Intent is mainly to gain understanding and to discuss the process. Process elements include objectives, business rules, activities, artifacts and roles.

3.4   Business Process Reengineering (BPR)

Business Process Reengineering is defined as "the fundamental rethinking and radical redesign of business processes to achieve dramatic improvements in critical, contemporary measures of performance, such as cost, quality, service, and speed" [Hammer 93]. BPR is thus concerned with completely rethinking how a service is provided or a product manufactured from scratch, without being constrained by current processes or organization.

As for EM, BPR is domain free. Any kind of business can be reengineered. The intent of using process technology is mainly understanding of the current process, and communication and analysis of the future process. Process elements are often just activities and roles.

3.5 Organizational Design (OD)

Organizational Design is the study of organizational performance using process models and computers. Given a business problem (e.g., "How will attendance to project meetings influence the product quality?"), a process model of a hypothetical organization is built. The expected performance of this organization is then predicted based on computer simulations on the model. OD makes extensive use of organizational theory in order to build models that result in accurate predictions. Two major efforts within this area are the Virtual Design Team project at University of Stanford [Christ 93] and the Process Handbook project at MIT [Malone 92].

OD is, as BPR and EM, domain free. The main intents of building process models are understanding and simulation. This requires more formal and detailed models than within e.g., EM and BPR. Process elements are activities, artifacts, roles, actors and tools. A notable difference from SE is the often detailed modeling of human actors, and their capabilities. The similarity to BPR is prominent, however, OD seems to put much more emphasis on computer simulation than the others.

3.6 Workflow Management (WM)

Workflow can be defined as "the sequence or steps used in business processes" [Marshak 91]. Marshak also requires that more than one person is involved in the process, and that the workflow consists of both sequential and parallel steps. *Workflow Management* is supporting and controlling the workflow.

An important objective in Workflow Management is to automatically route artifacts (documents) through a network to actors having predefined roles. The routing is done according to a set of predefined rules, and is often controlled by the state of the artifact (e.g., the price). This approach requires a relatively stable business process, as the rules are not meant to be changed on the fly.

The business domain of WM is information services, as for ISE. The intent of using process technology within WM equals the ones for SE, i.e., the full range from understanding to automation. Process elements in focus are business rules, activities, artifacts and roles. Business rules are particularly important for routing of artifacts.

3.7 Concurrent Engineering (CE)

The most frequently referenced definition of *Concurrent Engineering* is provided in [Winner 88]: "CE is a systematic approach to the integrated concurrent design of products and their related processes, including manufacturing and support. This approach is intended to cause the developers, from the outset, to consider all elements of the product life cycle from conception through disposal, including quality, cost, schedule, and user requirements."

The main idea of CE is to be able to run more activities in parallel with the aid of information technology and organizational restructuring. In addition, a life-cycle perspective on the product is encouraged, in order to reduce the overall costs. Main benefits of a successful implementation of CE are reduced time-to-market, improved product quality and increased productivity, leading to lower costs [Carter 92].

CE is in principle domain free. The intent is to focus on understanding, communication, and analysis of the current process in order to develop a new and improved one. The main process elements are activities, roles, and artifacts.

3.8 Computer Integrated Manufacturing (CIM)

Computer Integrated Manufacturing is defined by [Rembold 93] as follows: "CIM conveys the concept of a semi- or totally automated factory in which all processes leading to the manufacture of a product are integrated and controlled by computer."

The business domain of CIM is manufacturing. However, CIM focuses on processes that face *replication risks*, as opposed to SE, that face *design risks* [Bollinger 91]. The intent is simulation, planning, measurement, and automation. Main process elements are activities, artifacts and tools. A notable difference between CIM and other disciplines is the lack of focus on human actors.

4. Concluding Remarks

Our brief comparison of research areas shows that all fields have much in common with Software Engineering concerning application of process technology. The intents and process elements overlap more or less, even if the business domains differ.

One main finding is that the disciplines that are *most focused* in their business domain (like SE and CIM) have intentions of using process technology to a *further degree* than the disciplines that are domain free. They may also benefit the most.

Another main finding is that all areas intend to support applications of process technology that does not require *formal models*, while only a few support applications that do require formal models (SE, WM, CIM). Formal treatment of models may pose extra requirements, like model completeness and consistency.

SE may learn most from other disciplines concerning how to develop process models that are intuitively understandable for human beings (as focused by EM), and in a way that facilitate worker cooperation and motivation for organizational changes (like required for BPR). SE may also use more of the organizational theory that is the foundation of especially OD.

Particular to Software Engineering is the focus on metaprocess. It is obvious that this should be present in other areas as well, as reflection on the way process technology is applied is of utmost importance for improvement.

Acknowledgements

We would like to thank Guttom Sindre at IDT, NTH for his constructive critique, and valuable suggestions for improvements. Also thanks to Brian Warboys at University of Manchester and anonymous referees for helpful comments.

References

[Bollinger 91]   T. B. Bollinger, C. McGowan: A Critical Look at Software Capability Evaluations, *IEEE Software*, pp 25 - 41, July 1991.

[Carter 92]   D. E. Carter, B. S. Baker: *Concurrent Engineering - The Product Development Environment for the 1990's*, Addison-Wesley, Reading, Massachusetts, USA, 175 pages, 1992.

[Christ 93]   T. R. Christiansen: *Modeling Efficiency and Effectiveness of Coordination in Engineering Design Teams*, PhD thesis, CIFE, University of Stanford, California, USA, 317 pages, 1993.

[Curtis 92]   B. Curtis, M. I. Kellner, J. Over: Process Modeling, *Communications of the ACM*, pp 75-90, September 1992.

[Hammer 93]   M. Hammer, J. Champy: *Reenginerring the Corporation: A Manifesto for Business Revolution*, Nicholas Brealy Publishing, London, UK, 223 pages, 1993.

[Malone 92]   T. W. Malone, K. Crowston, J. Lee, B. Pentland: *Tools for Inventing Organizations - Toward a Handbook of Organizational Processes*, working paper, Center for Coordination Science, MIT, Massachusetts, USA, 21 pages, October 1992.

[Marshak 91]   R. T. Marshak: Perspectives on Workflow, In T. E. White, L. Fischer (ed.): *New Tools for New Times - The Workflow Paradigm*, Future Strategies, Inc., Alameda, CA, USA, pp 165-176, 1994.

[Naur 69]   P. Naur, B. Randell (ed.): *Software Engineering: A Report on a Conference sponsored by the NATO Science Comm.*, NATO, 1969.

[Promoter 94]   A. Finkelstein, J. Kramer, B. A. Nuseibeh: *Software Process Modelling and Technology*, Advanced Software Development Series, Research Studies Press/Wiley & Sons, 362 pages, 1994.

[Rembold 93]   U. Rembold, B. O. Nnaji, A. Storr: *Computer Integrated Manufacturing and Engineering*, Addison-Wesley, Wokingham, UK, 640 pages, 1993.

[Rumba 93]   J. Rumbaugh: Objects in the Constitution - Enterprise Modeling, *Journal on Object-Oriented Programming*, pp 18-24, January 1993.

[Sølvberg 93]   A. Sølvberg, D. C. Kung: *Information Systems Engineering - An Introduction*, Springer-Verlag, Berlin, Germany, 540 pages, 1993.

[Winner 88]   R. I. Winner et al.: *The Role of Concurrent Engineering in Weapons System Acquisition*, Report R-338, Institute for Defense Analyses, Alexandria, VA, USA, 1988.

Customising Software Process Models [*]

Graciela Pérez[1], Khaled El Emam[1] and Nazim H. Madhavji[1,2]

[1] School of Computer Science, McGill University
[2] Centre de Recherche Informatique de Montréal (CRIM)

1 Introduction

A core activity in the evolution of software process models is their customisation [7]. Customisation adapts a generic process model to make it more *effective* in the given software project.

Effectiveness of a customised process model[3] has two dimensions. The first dimension concerns *implementation*, which denotes the extent to which the process model is used and followed by practitioners in a particular project. The second dimension concerns *software project effectiveness*, which denotes the outcomes of the project (such as effort, cost, deliverables quality, predictability, etc.) if the process model is used and followed.

To ensure effectiveness, one important property of a project-specific process model is *congruence*. Congruence is a measure of how *fit* a process model is in the given development environment in which it is (intended to be) used. The development environment is referred to as the *process context*. A process model measures high on the congruence scale if its properties are suited to those of its process context.

Based on a review of the software engineering literature [8], process model congruence can be placed within a theoretical context. A basic theory of software process model congruence is depicted graphically in Figure 1. This theory predicts increased effectiveness with increased process model congruence. Accordingly, a customization method should strive to maximise the congruence property of software process models.

In this position paper we provide an *overview* of a method and accompanying tool that assist in the customisation of software process models. The method produces a congruence measure, and the tool operationalises the method and allows changes to the process model (i.e., performing what-if iterations) to increase its effectiveness. Further details of this work are available in [8].

2 Congruence Evaluation

In order to evaluate congruence, one must first develop a *contingency model*. A contingency model defines the relationships between the properties of a cus-

---

[*] This research was, in part, supported by NSERC, Canada, the IT Macroscope Project, managed by the DMR Group Inc., Canadian International Development Agency (C.I.D.A.), managed by WUSC, and Universidad de la Republica, Uruguay.

[3] In this paper we assume that the customised or project-specific process model is prescriptive rather than being descriptive.

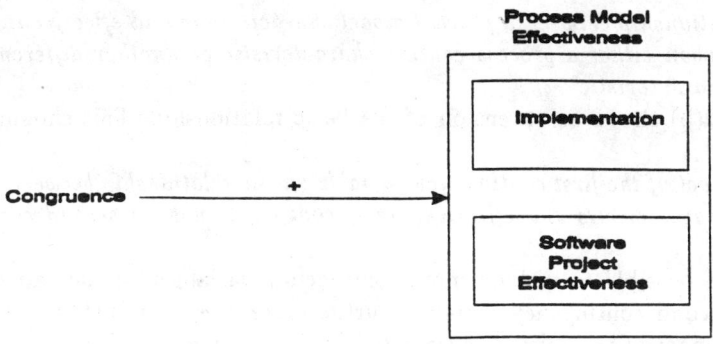

Fig. 1. A theory of process model congruence.

tomised process model and the properties of the process context. These relationships are the ones that affect the process model's effectiveness.

A contingency model represents variables and relationships as shown in Figure 2. There are always at least three variables in any single relationship: (a) a dependant variable, (b) an independent variable, and (c) a contingency variable.

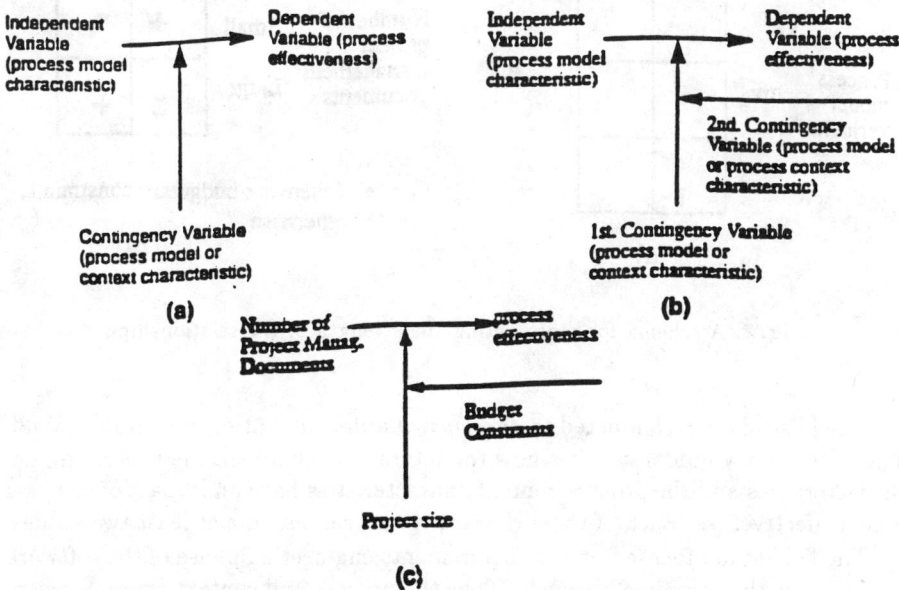

Fig. 2. A schema for representing the contingency model relationships

In a contingency model, the dependent variable is always *effectiveness* of a process model. The independent variable is always a process model characteristic. The contingency variable is either a process context characteristic or a process model characteristic. This is depicted in Figure 2(a). This diagram should be read as:

"the relationship between a process model characteristic and effectiveness is contingent upon either a process context characteristic or another different process model characteristic".

Figure 2(b) shows an extension of the basic relationship. This should be read as:

"the impact of the first contingency variable on the relationship between a process model characteristic and effectiveness is contingent upon a second contingency variable".

It is also possible to include more contingency variables, in the same manner as the second contingency variable. Furthermore, the relationship between any two contingency variables is represented in the same way as described in Figure 2(a). Figure 2(c) illustrates an example use of the relationships in Figure 2(b). Here, the figure states that: *"the impact of project size on the number of project management documents is contingent upon the budgetary constraints"*.

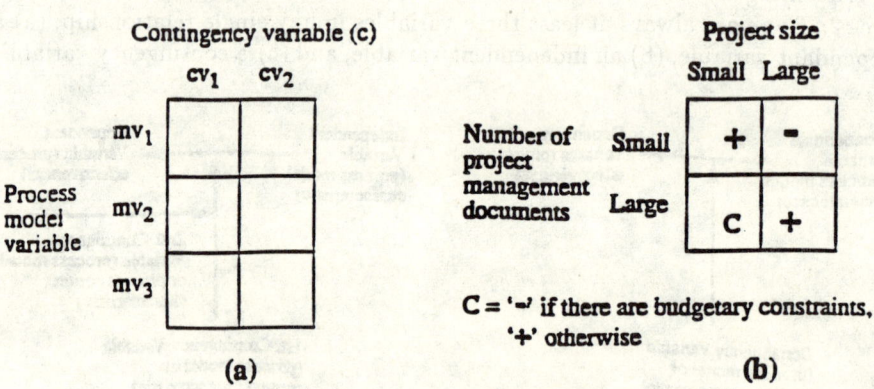

(a)   **(b)**

Fig. 3. A schema for representing the strength of the relationships

A field study was conducted to *develop* a qualitative contingency model [5] [8]. The contingency model specifies how the interactions between the process model characteristics and the process context characteristics have an impact on process model effectiveness. Each of these characteristics can assume at least two values.

The field study focused on the requirements engineering phase of the software process, not the complete life-cycle. Thus the process and context characteristics and relationships should not be generalised to the entire software process.

The projects studied were in organisations involved in the following businesses: retail, distribution, banking, insurance, manufacturing, government, services and telecommunications. The application domains were general business information systems: e.g., inventory control, decision support systems and payroll systems. The data gathering was performed through interviews, augmented with a number of survey questionnaires and analysis of project documents. In all, 33 interviews amounting 71 hours with 30 individuals (15 projects) were conducted. A detailed description of the empirical study and results is presented in [5].

Figure 2(a) shows a schema that can be used to represent the relationships in the contingency model, where the x axis represents a contingency variable and its possible values (e.g., cv_1, cv_2), and the y axis represent a process model variable and its possible values (e.g., mv_1, mv_2, mv_3). The relationship between the values of the variables can be represented using a "+", "−", "0" or a *blank*. A "+" denotes that the values of the process model variable and the contingency variable are likely to lead to increased process model congruence. A "−" denotes that the values of the variables are likely to lead to decreased process model congruence. A "0" denotes that the values of the variables are unlikely to affect process model congruence. A *blank* denotes that there is no relationship between the values of the variables.

At this stage of our research, the relationships are represented by "+", "−", "0" or a *blank* mainly because the evaluation is performed based on the qualitative contingency model described earlier, and that these are the type of relationships that we were able to identify in the field study.

The values of the variables in Figure 2(b) can be represented as shown in Figure 2 (b). For example, let us assume that the independent variable is the **number of project management documents**, the first contingency variable is **project size** and the second contingency variable is **budget constraints**. Then, Figure 2(b) describes that, in a given project, if the project size is small (large) and the number of documents produced are small then this would contribute to process model congruence (incongruence). If the project size is small and the number of documents produced is large then this is contingent upon the budgetary constraints (C). The large-large case would contribute to process model congruence because it is generally expected that a large project would produce a large number of documents.

Building on the above principles, a congruence evaluation method has been developed. To use the congruence evaluation method one must first assign values to a process model's and a process context's attributes. This could be done by a process engineer familiar with the process model and the process context (the issue of reliability of these assignments is discussed in the conclusions). A congruence measure is then computed based on the inter-relationships amongst the process model and process context attributes. In other words, *the strength of the individual relationships are aggregated to produce the overall congruence.* This method is described in detail in [8]. The outputs produced by the method are:

- **Congruence index :** It is a real number between -1 and 1 that evaluates the congruence of the process model with its context. A value of 1 indicates the best possible match, and -1 the worst.
- **Congruence values of each process model attribute:** They provide

a congruence measure of each process model attribute. This permits the determination of the process model attributes which are or are not congruent with the context.

- **Congruence values of each context attribute:** They provide a congruence measure of each process context attribute. This permits the determination of the process context attributes with are or are not congruent with the process model.
- **Trouble-spot list** This is a list of all the relationships, where a "−" or "0" value exist for the process model-context attributes under study.

To validate our approach to congruence measurement, another empirical study was conducted to demonstrate that the congruence measures (i.e., congruence of each attribute as well as the overall congruence index) produced by the evaluation method characterise congruence. Specifically, the validation follows the approach described by Fenton [6] whereby empirical evidence is gathered to show a mapping between an empirical relation system and a numeric relation system. This study shows that the congruence measures reflect the intuitive notion of congruence held by senior software process professionals[4]. Details of the study and the results can be found in [8].

The congruence evaluation method has been operationalized in the form of an automated tool. In addition, the tool enables cycles of changing one or more attribute values and re-evaluation of congruence, thus facilitating experimentation with process model customisation. Details of this tool and its functions[5] can be found in [8]. Furthermore, the evaluation method and tool have been employed in a number of case study applications. An example application is given below.

3   Process Model Customisation

The applications of the method and tool show how the evaluation results can be used to provide assistance in: process model customisation, process model design, the selection of process models, and in process reuse [8]. Here, however, we demonstrate through an example the application of the congruence evaluation method for process model customisation only.

The first two steps of the evaluation method consist of characterising the process and the context under study. Figures 3 and 3 show the characterisation of one of the projects we studied. The process model and context information is input to the tool. Following this, the evaluation algorithm is performed and the outputs are displayed, as shown in Figure 3. The screen dump shows only a part of the information, but it is possible to scroll through the whole list of attributes-congruence and trouble-spot list.

---

[4] The mentioned study assumes that a congruence measure is at most on an ordinal scale.

[5] Another function of this tool is process model design assistance. With this, a process engineer inputs the characteristics of a process context, and the tool assists him/her in designing the *most fit* process model.

Process Model attributes	Value	Process Model attributes	Value
Exit criteria	Lower than Stand.	Existence of templates	Few
Output/Deliv. divisionalisation	Yes	Existence of examples	Many
Integration mechanisms	Low	Effort examining current system	High
Devices for underst./valid. function.	Prototypes	Age of method	Old
Coordination mechanism	Stand. of deliv.	Type of modeling in PA	Functional+concep.
Formalis. of change management	Low	COTS orientation	COTS is part
Resources for PM. control & plann.	High	Workflow depiction in method	Not Obvious
Amount of Project Manag. docum.	Large	Role specialisation in model	High
Emphasis on control and planning	High	Relative difference of model	Different
User s view formalism	Simple	Incremental implementation	Yes
Size of user's view of model	Large		

Fig. 4. Characterisation of Process model 1

Context attribute	Value	Context attribute	Value
Accuracy of estimates required	High	Package modification needed	No
User experience with method	Low	Use of consultants (process)	Extensive
Delegation of author. to users	SC validates	Availability of users	Available
Manag. method for personnel selec.	By capability	User wants a package	Yes
Ability to change	High	Package already purch./selected	No
IS spending	High	Technology strategy	Proc. improv.
Management expectation	Patent	Highly interactive system	No
Project size	Large	Time of package decision	Early
Application uncertainty	Low	Existence of a legacy system	
Application complexity	Low	User wants something new	No
Number of user departments involved	Few	Style of project manager	Strict control
Budgetary constraints	No	Centralis. of decision making	Centralised
User interface	Not critical	Schedule constraints	No
Prototyping tool support	Not Available	Management Method	Deliv. oriented
Main user emphasis	Functionality	Use of consultants (any cons.)	Extensive
Users IS experience	High	Information system size	Large
Users communication skills	Low	Reward system	Long term
Customs. of business process	Low	PM technical involvement	Low
Scope of system	Narrow	Extent of tech. driven culture	Low
Process experience of practitioners	Large	Stand-alone package	Yes
Organisation has own standards	No	User involvement	Low

Fig. 5. Characterisation of Context 1

In this example, the evaluation yields a congruence index of 0.71. In order to analyse how to improve congruence, and hence to customise the model further, we can search through the list of process attributes and, for the ones that have low congruence values, we should check in the trouble-spot list to determine the reasons. For instance, Devices for underst./validation = Prototypes and Integration mechanisms = Low have congruence values −1 and 0, respectively. The trouble-spot list shows that the reason for these low values is that the following relationships:

(i) Devices for underst./validation = Prototypes
and Application uncertainty = Low,

```
01/09/93            McGill - CONGRUENCE EVALUATION SYSTEM           2:04:10 pm
```

PROCESS MODEL CONGRUENCE

PROCESS MODEL CHARACTERISTICS - CONGRUENCY

	Process Characteristic	Value	Congruence
CONGRUENCY INDEX: 0.71	Exit criteria	Lower than	0.00
	Output/Deliv divisionalisation	Yes	1.00
	Integration mechanisms	Low	0.00
	Devices for underst/valid func.	Prototypes	-1.00
	Coordination mechanism	Stand. of	1.00
	Formalis. of change management	Low	1.00
	Resources PM control & plann.	High	1.00
	Amount of Project Manag. daily	High	1.00
	Emphasis on control and plann.	High	1.00

PROCESS MODEL - TROUBLE SPOTS

Rel	Process Characteristic	Value	Related Characteristic	Value	Type
-	Integration mechanisms	Low	Project size	Large	C
-	Devices for underst/valid	Prototy	Application uncertainty	Low	C
-	Devices for underst/valid	Prototy	Prototyping tool support	Not Ava	C

Fig. 6. Output generated by the evaluation method when input was Process model 1 and Context 1

(ii) `Devices for underst./validation = Prototypes` and `Prototyping tool support = Not Available`, and
(iii) `Integration mechanisms = Low` and `Project size = Large`

are related negatively. If we would like to determine the impact on overall congruence, of changing some of the process characteristics, we can do so by modifying the values of only the desired attributes. The result of re-evaluating the congruence after changing the values of `Devices for underst./validation` from `Prototypes` to `Models` and `Integration mechanisms` from `Low` to `High`, are the following: the overall congruence index and the congruence of `Devices for underst./validation` have increased (to 0.86 and 1 respectively), but the congruence of `Integration mechanisms` has not. This is due to the fact that, now, `Integration mechanisms = High` and `Application uncertainty = Low` are related negatively. From these results, we can conclude that in this particular context there are no benefits, with respect to congruence, of changing the value of `Integration mechanisms` from `Low` to `High` because the congruence is not increased by this change.

What this small, but significant, example shows is that during customisation, one can experiment iteratively with the desired process and context attribute values. Of course, such cycles should be carried out in an organised customisation process (which is not intended to be discussed in this paper). The general goal of such experimentation is to increase the congruence of the process model, and hence achieve a higher degree of customisation. The role of the method and tool is thus to support decision making during the customisation process.

4 Comparison With Related Work

An important dimension for comparison is the level of granularity of the process information treated during customisation. Boehm and Belz, in [2], propose a process model generator as a technique to select life-cycle models based on process drivers. Our approach, on the other hand, focuses on detailed attributes of a process model and on evaluating congruence based on the values of these attributes. A similar comparison can be made between our work and that of Davis et al. [3] because they too address process models at the level of general life-cycle models.

In a paper by Alexander and Davis [1], the authors describe a method for selecting life-cycle models based on a set of selection criteria, which we consider to be more detailed than those of Boehm and Belz. In contrast, we do not prescribe a specific type of life-cycle model for a given project. Rather, we identify a set of properties that a detailed process model should possess, given a particular set of context attributes. A process engineer can then build an appropriate process model based on the identified process attributes. In addition, our approach evaluates the congruence of a given process model. This evaluation cannot only be used to select process models but it can also be used to customise and improve a process model. Furthermore, we have implemented a tool based on our methods.

5 Conclusions

The problem of customising software process models is recognised to be of considerable importance in software engineering. This paper has described a method and tool for evaluating the congruence of a software process model with a given environment, and for applying the congruence measure during customisation. Thus, this research can be considered a major step towards solving the general problem of process engineering.

Suggested future research should focus on providing further validation of the overall customization approach described here. Specifically, two aspects require attention. The first aspect concerns inter-rater reliability. The second aspect concerns validating the process model congruence theory.

In order to use the congruence evaluation method and system described in this paper, one has to assign values to the process model and process context attributes. An inter-rater reliability study would investigate the consistency and repeatability of this assignment across different individuals. The results of such a study would provide directions for increasing the reliability of congruence measurement.

A quantitative empirical validation of the process model congruence theory would raise our confidence that customization methods ought to increase the congruence property of project-specific process models. Testing this theory is now more plausible given that a congruence measure has been developed. However, measures of process model effectiveness need to also be developed before theory testing can commence. In the context of the requirements engineering process, a requirements engineering success instrument has been reported [4]. This can be used to measure one dimension of process model effectiveness.

References

1. L. C. Alexander and A. M. Davis. Criteria for selecting software process models. *Proceedings of the 15th International IEEE COMPSAC*, pages 521–528, 1991.
2. B. Boehm and F. Belz. Experience with the spiral model as a process model generator. In *Proc. 5th Int. Workshop on the Software Process, ACM SIGSOFT*, Kennebunkport, Maine, USA, October 1989. ACM Press.
3. A. Davis, E. Bersoff, and E. Comer. A strategy for comparing alternative software development life cycle models. *IEEE Transactions On Software Engineering*, 14(10), 1988.
4. K. El Emam and N. H. Madhavji. Measuring the success of requirements engineering processes. In *Proceedings of the Second IEEE International Syposium on Requirements Engineering*, 1995.
5. K. El Emam and N.H. Madhavji. A model of the factors affecting the success of a requirements engineering process. Technical report, Macroscope Project Phase III Report, Montréal, February 1994.
6. Norman E. Fenton. *SOFTWARE METRICS A Rigorous Approach*. Chapman & Hall, 1991.
7. N.H. Madhavji, K. El Emam, and T. Bruckhaus. An on-going study of factors causing process evolution. In *Proceedings of the International Workshop on the Evolution of Software Processes*, Gault Estate, Mt. St. Hilare, PQ., Canada, January 1993. McGill University and rMise.
8. G. Perez. A system for evaluating the congruence of software process models. Master's thesis, McGill University, 1994.

Process differentiation and integration: the key to just-in-time in product development

J. Henk Obbink

Philips Research Laboratories
Prof. Holstlaan 4, 5656 AA Eindhoven, The Netherlands
tel. +31 40 742575, Fax +31 40 744004
internet: obbink@prl.philips.nl

Abstract. In this paper it is argued that in order to achieve the required reduction in product lead-time for embedded software a proper differentiation and integration of a number of key developemt processes like: usability engineering, hardware engineering and software engineering should take place. It is furthermore argued that this makes only sense if an appropriate modular system architecture exists, which admits a flexible Δ development approach for its major components.

1 introduction

Industrial organisations are continuously striving to improve their product development capabilities. A capability is defined as the ability of a development organisation to consistently deliver a product or an enhancement to an existing product:

- that meets customer expectations/needs
- with minimal defects
- for the lowest lifecycle costs, and
- in the shortest time

On many markets, for example the market for consumer electronic products like: TVs, HiFi-set, mobile phones, etc, the economic product life is becoming shorter and shorter. Products become more rapidly obsolete because the customer is willing to pay for the newest, innovative products. Personal computers are another notable example of this phenomenon in the computer domain. As a result, the importance of timely product development as a 'competitive weapon' of a company is becoming increasingly important. The company must become a *time-based competitor* [1]. Hitting the market at the right time becomes critical for making profit or even survival. These modern electronic products typically include: hardware (mechanics and electronics), software, documentation, training and support services. The tough competition in the world markets will soon lead to the situation that for instance the product lead-time for a typical consumer product will have to be reduced from 1,5 years to about 0,5 years. Already significant reductions in the throughput time have been achieved in the back-end of the industrial process e.g., manufacturing, distribution and sales. Also the back-end of the development process itself has been largely automated. To further reduce the lead-time the conception and front-end development times have to be reduced significantly. To achieve this strong reduction in the development lead-time requires a judicious and meticulous integration and reengineering of the software-engineering, hardware-engineering and usability-engineering development processes in order to achieve the capability improvements, w.r.t. the significant lead-time reduction. The corner stone for enabling this substantial decrease in product lead-time, leaving the rest of the required product qualities untouched, lies in the integration of the above mentioned engineering processes around a domain-specific product **reference architecture** and a so-called Δ development process. A typical example of such a product system architecture can be found in the SPRINT method [7], [9]. The SPRINT method has been developed within Philips Electronics and integrates formal Specification-, Prototyping- and Reuse technologies. The SPRINT method contains as its heart a combined SW/HW domain component model, fitting in an overall HW/SW system architecture. The SPRINT component model is shown in Figure 1. In this figure we on the top the generic model of the component. The electrical signals are modelled as horizontal dashed lines. The other interfaces of the component are dashed lines for observers and solid lines for transformers. Arrows indicate the direction of *initiative*. A step is a represented as a command of the component itself. In the bottom part of the figure the box diagram of a simple tuner component is shown.

In SPRINT the components are configured in various layers of the SPRINT architectural model which is shown in Figure 2. Each of the layers is a higher aggregated component itself. When a new product is required changes can occur in all three main layers. The top control system layer is the most user interface specific and the Δ's in the Control System are therefore mostly effected by the outcome of the usability engineering process.

The Δ's in the Hardware Specific layer are mostly effected by the outcome of the HW/SW codesign process.

The Δ's in the layer of reusable domain component is usually small.

The overall process management of a so-called Just-In-Time Δ development process (or process differentiation) now admits only those Δ's that are compatible with the lead-time requirements on the overall product development process. In order to achieve this a tight integration of the: usability engineering, Embedded Software Engineering and HW/SW codesign processes. This means that a software only view is simply not sufficient anymore. A number of disciplines and cultures have to be integrated in a seamless fashion around the generic product architecture. Most of these processes are partly understood and process technology could be of great help in clarifying the inter- and intra-process integration and automation issues [8].

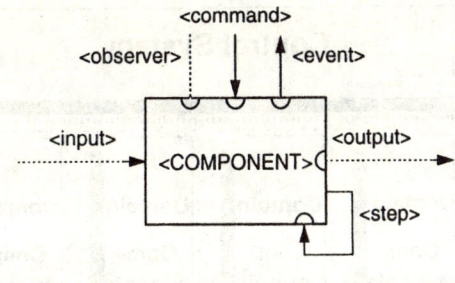

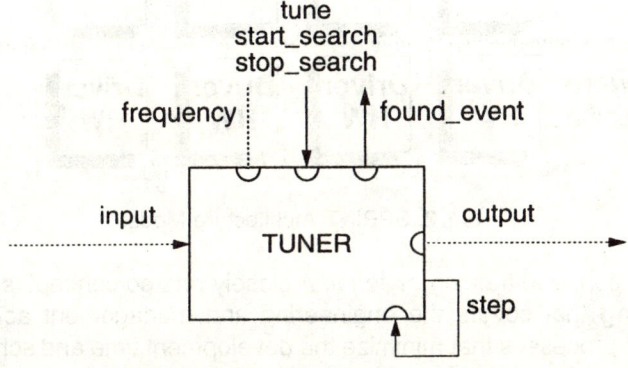

Fig. 1. SPRINT Component Model

The following notions have been used or will be used in the rest of the paper:

- Component
- Architecture
- Co-engineering
- Concurrent Engineering
- Usability Engineering

A **Component** is a self contained entity containing hardware and/or software that provides coherent and reusable functionality at the software level. It realises a maximal abstraction on the underlying hardware (if any). The components are the basic building blocks of the SPRINT system architecture. With **Architecture** we mean the set of components and their relations [4], [5], [10]. Already in the old days building architecture used similar definitions. In Vitrivius's "The Ten Books on Architecture", the terms order, arrangement and economy [12] are introduced. In computers and software the first of the architectural dimensions of a system is the view of the system presented to the one who uses the system [17], the second dimension is the internal organisation of the system, the third dimension addresses the implementation guidelines [18].

Co-Engineering covers the engineering and management activities, techniques and processes that facilitate the co-development of subsystems of a different nature e.g., electronic hardware, mechanical hardware, silicon, software, user

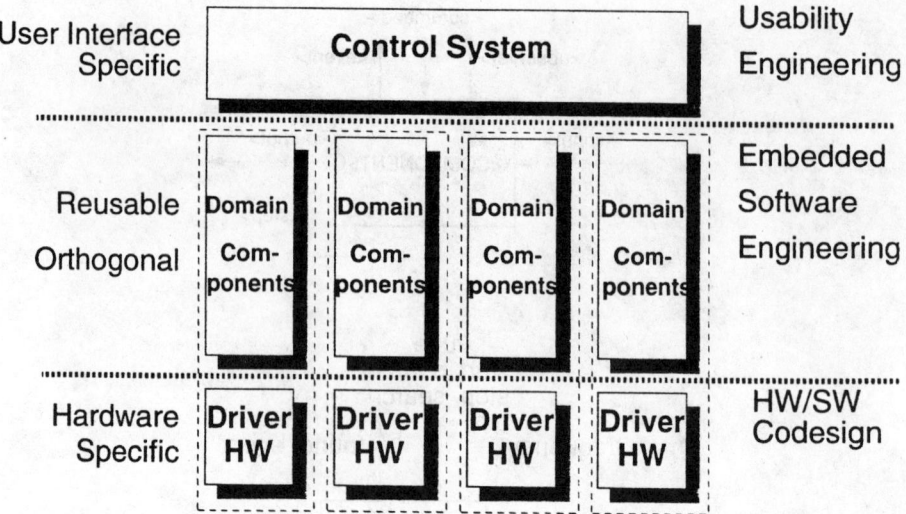

Fig. 2. SPRINT Architecture Model

documentation, and training material. A closely related concept is **Concurrent Engineering** that covers the engineering and management activities, techniques and processes that minimize the development time and schedule(cycle-time) of a product; this is achieved through optimization of the concurrency in the performance of product development tasks (e.g. specifications, design, code), and minimization of inter-organizational/ functional communication through multifunctional teams. **Usability Engineering** is defined as the engineering and management activities, techniques and processes that optimizes the usability of a product (e.g., minimization of user training, minimization of potential operator errors, optimization of time needed to perform the most used functions) [2].

2   Types of Product Creation Processes in Industry

I will take the example of the electronics industry were we can identify a number of so-called Product Creation Processes (PCP). Each process is organised in such away that a major concern, or risk is managed. Furthermore, each type of PCP is characteristic of a particular stage in the life-time of a system. These PCPs differ in at least three main aspects: newness of the domain (Domain Δ), newness of the functionality (Functional Δ) and newness of the technology, both process-technology Δ's and Product-technology Δ's. For electronics we have the following five types of PCPs:

I A PCP which enables the variation **WITHIN** product families:
  The families are characterised by a known domain and known functionality. Admissible variation is restricted to User Interface style, appearance and dynamics and selection of already implemented functionality. (Stroke

versions.) The key issue is lead-time reduction, achieved through architecture based reuse and automatic code generation. Software development concentrates on the development of the User Interface and the linkage of pre-implemented basic functionality in a well-prepared domain specific architecture using a strict Δ approach. For the layer of Domain Components only negative $Delta$'s are allowed. Software development concentrates on development WITH reuse. The key ingredients are automation of a large part of the process and no iteration is allowed. This type of PCP is useful for quite mature products, with stable architectures, in high volume markets and known processes. Think about portable audio cassette recorders, electric shavers. The domain is known, the functionality is known and the implementation technology is known. So there are no significant barriers to predict the development time adequately. The key issue is the integration of usability engineering process with the embedded software engineering process in the architectural framework of Figure 2.

II A PCP which enables the variation **ON** product families:
The families are within a known domain and the variation is that of new functionality (New at the time the family was conceived and developed). An example is for instance the addition of RDS functionality to a standard audio set. Software development now concentrates on the evolution of the existing architecture and the development of the new functionality, the adaptation of the existing components, etc. For the unaffected parts of the architecture the previous PCP should be followed. Software design concentrates on design WITH reuse and a little bit on design FOR reuse. Depending on the exact lead-time requirements, only Δ's that are complient with those requirements can be accepted. The new functionality is a positive $Delta$ on the middle- and optionally the bottom-layer in Figure 2. If the bottom layer is changed than the key issue of lead-time reduction becomes the integration of HW/SW codesign in the small(Only key-components) with the embedded software engineering process.

III A PCP supporting the development **FOR** new families in a known domain:
E.g. a new generation of a certain product family is needed. The main reason for change is that new technologies have become available in either hardware and/or software that the development of new family becomes more cost-effective than the existing family. Development now comprises the whole process: i.e. domain engineering, requirements-specification, architecture, design, implementation, prototyping, HW/SW codesign, etc. Software development now concentrates on design FOR reuse. The main purpose is to devise such a process that the development processes of type I and II are become possible. Experience has taught that it requires a number of iterations before a sufficient level of systematic reuse can be achieved. This means before the most appropriate process- and product architecture has been developed. The key issues are now the overall process- and architecture-engineering. In this case also the overall HW/SW architecture must be developed. This is an exmample of HW/SW codesign

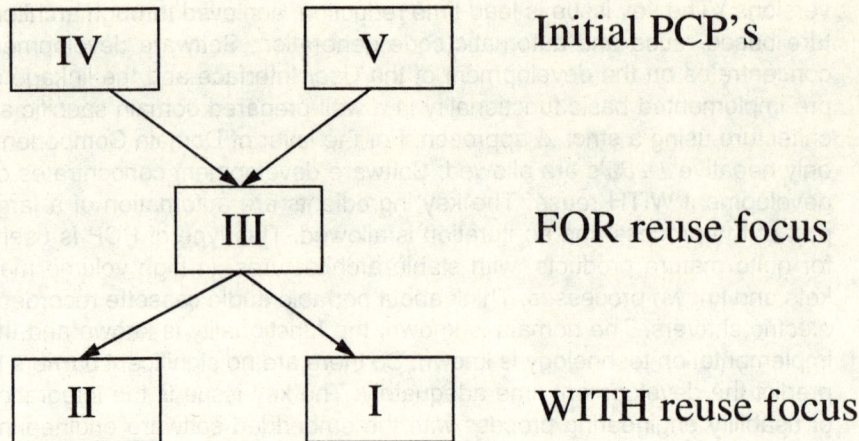

Fig. 3. PCP Evolution Process

It is tempting to identify these different types of PCPs with the five maturity levels of Watts Humphrey [16] in reverse order. Type I PCP would then correspond with the optimized process. Type V would than correspond with an initial process. Further analysis has to be performed to investigate that relationship. Throughout the discussion of the above mentioned PCP types the notion of: process, architecture and product families have been used. We refer to [19] where the process of composing product families from reusable components is described. In [20] the relation of architecture and the development process is explained in more detail. Both articles deal with the application of similar ideas in the context of communication systems.

In addition to the mentioned types of PCP we can distinguish the following types of products in the electronics industry:

- *closed systems*, with so-called embedded software. Think of a classical measuring instrument, or a shaver with motor and battery control, a brake controller in a car a micro-wave oven controller. The main functionality is often provided by the hardware.
- *open systems*, with so-called embedded hardware. Think of the stand-alone PC, CD-i, a cartridge game-boy. The software, which carries the main functionality, is unbundled from the hardware. The software is written according a standard, either a public or a proprietory application programmers interface.
- distributed embedded systems. These systems are an extension of the previous type when networking and distribution is introduced. Think of interactive TV or tele-CD-i.

If we combine these three product types with the five different PCPs we get fifteen different processes that have to be managed and developed. A major issue is the management of the technology transfer process, when a certain product migrates from one process type to the other. Especially, when in parallel the product type moves from one type to the other.

In the following sections I will give a short description of the HW/SW codesign and usability engineering processes and issues. Both processes have great potential for lead-time reduction of the overall development process. For a more elaborate description than give here I refer to [6] for HW/SW codesign and [13] for usability engineering.

3 Co-Engineering

A special branch of Co-engineering in electronics is HW/SW-codesign. Codesign is often referred to as the "integrated design of systems implemented using both hardware and software components" [11]. Hardware/software (HW/SW) codesign is predominant in the engineering of tightly coupled systems with hardware and software modules interacting to solve a certain task. Such systems are not new. However, methodologies that currently apply and trade off design techniques from both spectra are only now emerging [6]. The goals of this design technique are optimised functionality of heterogeneous systems and reduced development time. To accomplish these goals, system modelling techniques must not presuppose an implementation of various components in hardware or software. Subsequent design refinement of hardware and software components should be performed concurrently, which requires appropriate coordination and consistency control methods. Furthermore, mixed hardware/software specification, modelling, simulation and verification techniques should be available to the designer. The key to a unified codesign process is the concurrent pursuit of software and hardware threads in the design process. Decisions made in one thread can significantly affect the other. Therefore, planning, scheduling, and synchronization of design activities should be ensured through well-defined task, flow, and design data management techniques provided by a suitable framework.

In a traditional design strategy, the hardware and software partitioning decisions are fixed at an early stage in the development cycle and the hardware and software design are developed separately from then onward. With advancements in technology, however, it becomes possible to obtain special-purpose hardware components (ASICs) at reasonable costs and development time. Some designs also call for some programmability in the end product. This suggest a more flexible design strategy, where the hardware and the software designs proceed in parallel, with feedback and interaction occurring between the two as the design progresses. The final hardware/software split can then be made after the evaluation of alternative structures with respect to performance, programmability, non recurring (development) costs, recurring (manufacturing) costs, reliability, maintenance, and evolution of design. This design philosophy helps reduce the time to market, which is in the case of tough competitive markets equivalent to time to money. A few months too late on the market can make a difference between profit or loss. The interaction between the different design domains must be better understood and modelled. The common waterfall model for the process flow will not be adequate. With a high degree of interaction between the design domains and abstraction levels, the pure top-down design flow cannot not be maintained. Instead, the process structure must allow for a more irregular flow between design domains and abstraction levels. The process structure must then express the information states of the design rather than being a task flow description, since the design may be carried out at various abstraction levels and in different design domains simultaneously. This will also require a tighter integration of the CASE and CAD tools.

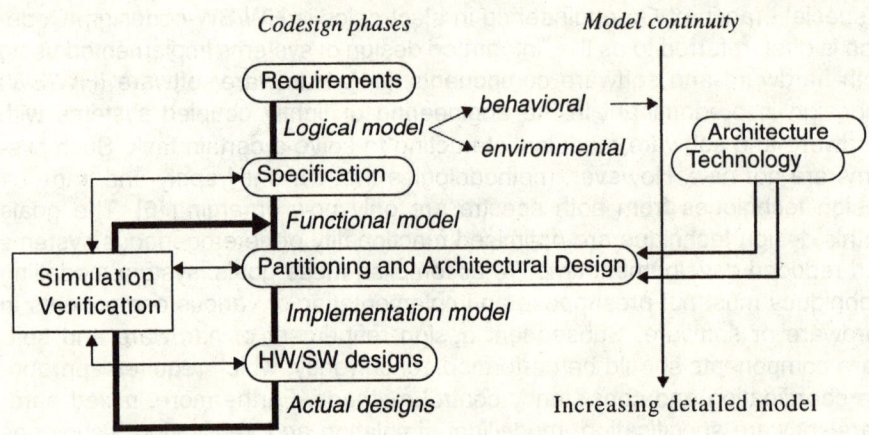

Fig. 4. HW/SW Codesign Process

We have a number of stages:

- Analysis of Constraints and *Requirements*.
  In this step, the basic system's characteristics are defined based on the users's needs and customer's specifications. Project objectives, constraints, and requirements as stipulated by the client often lack coherence and completeness. Typical design facets captured in the analysis step includes marketability based on study of consumer expectations, real-time performance requirements, realisation technology, programmability, power consumption, product size, nonrecurring costs (development), recurring costs (manufacturing), environment in which the product is to be used, reliability, maintenance, design evolution, and recycling costs.

– System *Specification.*
The system specification is the result of the analysis step. It is a formalised specification suitable for electronic processing from which a designer derives and develops modelling algorithms. Such algorithms may be simulated using rapid prototyping tools based on state charts or queuing models for a performance-oriented simulation. The simulation is, in fact, the first opportunity for system designers to demonstrate the idea to both management and marketing staff.

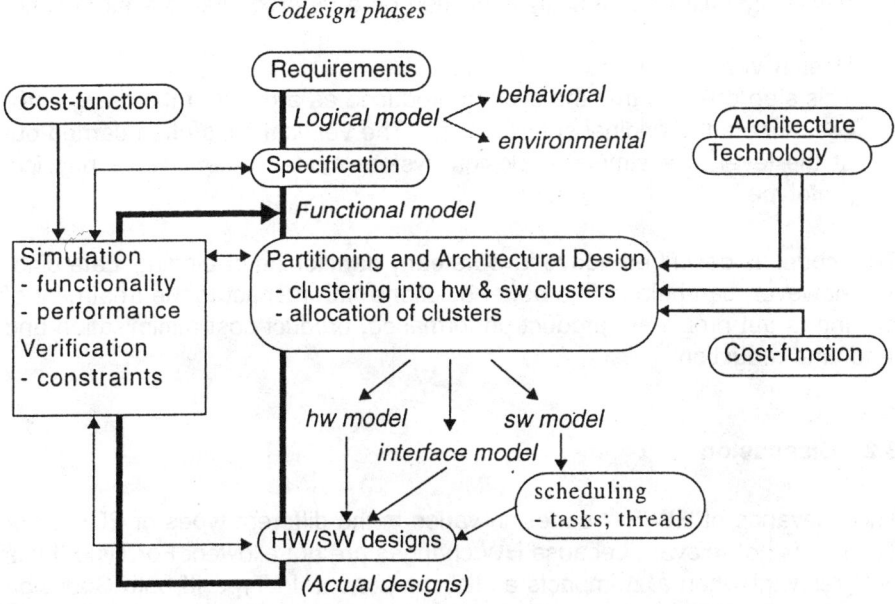

Fig. 5. Partitioning Issues in HW/SW codesign

– Hardware and Software Design

- Hardware Synthesis and Configuration.
  In this phase a hardware platform that executes the code produced in the software generation step is designed. It is constructed from the hardware description obtained during partitioning. The synthesis involves technology binding by translation of HW descriptions into physical units. Software components depend on decisions made after partitioning and, therefore, the software generation step is dependent on the decisions made in the early phases of the hardware synthesis.
- Software Generation and Parametrisation.
  In this phase, software modules are generated for the synthesised and configured hardware (here an opposite view is taken; hardware is assumed to be generated prior to the software). Proper interaction of the

components of the hardware and software is ensured by a scheduler that is either implemented in hardware, software or both.

- Interface Synthesis
  Interface Synthesis provides a means of hardware and software synchronisation. Typically, signal exchange (HW), semaphore(SW), or interrupt driven schemes are employed in this phase.

- Hardware/Software Integration and Cosimulation
  The problem of HW/SW integration and cosimulation is dual to partitioning and is equally difficult. It leads to a working prototype that can either be physically built or eventually exercised on a heterogeneous system simulator.

- Design Verification
  This step ensures that the system produced as a result of the design process meets the original specifications. The verification is often carried out at the level of a simulated design system model rather than a physical prototype.

The model as described above favours early partitioning or binding. Late binding, however, can help to find better solutions with respect to the treatment of moving target problems, product performance, product cost minimisation and lead-time reduction.

3.2 Discussion

The relevance of HW/SW codesign varies in the different types of PCPs. For Type I it is not relevant because HW changes are not allowed. For Type II it is only relevant when a Δ impacts a HW-component. In Type III both Codesign in the small(CITS) and Codesign in the large (CILT) are relevant. In Type IV and V it is relevant.

4   Usability Engineering

Usability engineering is the next key process that can contribute significantly to the required lead-time reduction. Usability can be thought of a cumulative attribute of a product. When a product development team designs a product, it tries to include the features people need to accomplish tasks, present those features in a manner that people intuitively grasp and find efficient to use in the long term. They also attempt to eliminate the potential for critical, design-induced mistakes and to include design qualities that make people feel good about using the product. Almost every aspect of a product's appearance, feature set, and interaction scheme choreography, if you will) affects these goals. When designing a computer based product hundreds or even thousand of detailed design considerations have to be represented. Total or overall usability depends on how well design teams handle all these details. If a product is to have a high level of usability, the product development team must approach the design of the user interface–the elements of the product that people interact with–at a broader level conceptual level as well. Often, products noted for their usability may in fact be mechanically or electronically complex. This apparent paradox is often the result of good work of usability professionals, whose goal has been to evoke in users a simple mental model (framework) of how the product works and to protect them from learning unnecessary design details. Usability may also be thought of as a design philosophy that places users' needs high– if not– first on the list of design priorities. Obvious usability benefits:

- reduced customer support and service costs
- avoidance of costly delays in the product development schedule in order to fix usability problems before going to market
- reduced customer training costs
- simpler-to-prepare product documentation
- accurate, ready-to-use marketing claims based on tests

A simplified view of the usability process, consisting of 6 steps, is shown in Figure 6.

1. Conduct User Research
   The users of the product under development are researched. This activity may be limited to talking informally to a number of people, or may go as far as conducting a set of focus groups, conducting a questionnaire survey, or observing and interviewing people as they use similar products. Often companies perform detailed analyses of user tasks a basis for understanding control and display requirements or information flow (I/O) requirements. These activities lead naturally to the next task.
2. Set Usability Goals
   The concept of setting a usability goal–taking a quantitative view of usability-– is new to some. However, the exercise forces designers to think specifically about how users will react emotionally to a product and how well they will be able to perform the required tasks. Furthermore, designers can set goals how high people will rate a product's intuitiveness or how enjoyable it is to use. The process is the same as setting a goal for the gross

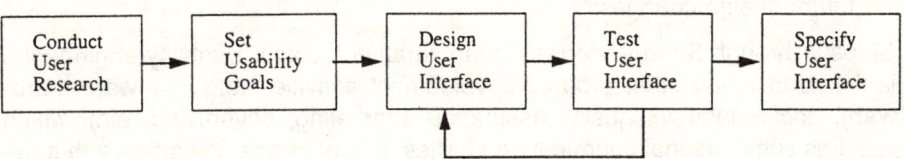

Fig. 6. Basic User Centered Design Process

weight or power consumption of portable computer, except that designers are preparing to measure usability. They can base goals on the performance of competing products, a company's current product, or their best judgement of user's needs and the potential of new designs.

3. Design the User Interface
Usability goals, coupled with the results of user studies, help designers formulate alternative concepts for the user interface.

4. Model the User Interface
Many user interface design team develop several alternative concepts for the user interface in parallel, describing them in the form of interaction sequence diagrams, paper-based screenplays or electronic prototypes(for software), or physical mockups that people can actually touch(for hardware). Typically, the associated design effort is guided by a product-specific or corporate-specific set of design principles drawn from the human factors literature and experience.

5. Test the User Interface
Once there is something to show, users can get involved again by participating in focus group discussions of design alternatives or usability tests that provide feedback for designers. Normally the feedback comes at two levels: opinions about the overall design and detailed suggestions for improving specific design elements. Interactive prototypes of varying degrees of realism also enable designers to observe usability at both general and specific level. Designers can use this feedback as a basis for choosing a preferred conceptual model and associated solution an then refine it. Usability specialists can also bring electronic prototypes or working models of near final designs together with potential users, either in the laboratories or the field, to validate design features and detect any remaining usability problems that need to be fixed. Sometimes this testing process is extended to include field testing of production-line products as a means to document the benefits of a user-centered design process.

6. Specify the User Interface
The usability endgane involves specifying the user interface so that it can be implemented properly by programmers, engineers, and the like. Depending on corporate documentation requirements, usability specialist may do any of the following: write detailed specifications, build a computer-based prototype or mockup that reflects final design changes, create menu hierarchy maps or state diagrams that illustrate the user interface logic, or create templates and style guides to ensure the consistency and quality of future design changes.

Steps 3 through 5 are performed in an iterative fashion. Usability engineering is linked to many other product development activities (e.g., software, hardware, documentation, quality assurance, marketing, advertising, etc), which requires substantional coordination abilities. It is of utmost importance that usability engineering is injected early in the product creations process. Nothing is more frustrating than polishing up a nearly finished design, rather than getting involved earlier on when major user interface decisions are made [13], [14].

4.1 Discussion

Usability engineering is important for all PCPs, but the biggest challenges ly in the ability of an organisation the integrate usability early in the PCP in order to avoid all kinds of costly iterations.

5 Acknowledgements

I would like to thank Hans Jonkers, Rob van Ommering and Jürgen Müller about architectural discussions, Ernst-Jan van Nigtevecht about HW/SW code-sign, John de Vet about Usability Engineering, René van Loon about Just-in-Time Development, Peter van den Hamer about product families, Jan Rooij-mans and Hans Aerts about Software Process Improvements. Ben te Boekhorst and Wiecher van Vugt about Δ development.

References

1. G. Stalk jr., T. Hout. *Competing against Time*. The Free Press (MacMillan), New York; 1990, ISBN no: 0-020915291-7.
2. *Trillium: Model for Telecom Product Development & Support Process Capability*. Internet edition, Bell Canada, Release 3.0 December 1994.
3. E. Gamma, R. Helm, R. Johnson and J. Vlissides. *Design Patters: Elements of Reusable Object-Oriented Software*. Addison-Wesley, 1995, ISBN no: 0-201-63361-2.
4. B.I. Wit, F.T. Baker, E.W. Merritt. *Software Architecture and Design: Principles, Models and Methods* Van Nostrand Reinhold, 1994, ISBN no: 0-442-01556-9.
5. R. Allen, D. Garlan. *Formalizing Architectural Connection*. In *Proceedings ICSE 16*, May 16-12,Sorrento, Italy, pages 71–80.
6. J. Rozenblit, K. Buchenrieder. *Codesign: Computer-aided software/Hardware Engineering*. IEEE Press, 1995, ISBN 0-7803-1049-7.
7. H.B.M. Jonkers. *An overview of the SPRINT method*. In: Woodcock, Larsen (Eds.). FME'93: industrial-strength formal-methods, Springer-Verlag, LNCS 670, pp. 403-427, 1993.
8. J.H. Obbink. *Systems Engineering Environments of ATMOSPHERE*. In: Endres, Weber (Eds.). Software Development Environments and CASE technology, Springer-Verlag, LNCS 509, pp. 1-17, 1991.
9. L.M.G. Feijs, H.B.M. Jonkers and C.A. Middelburg. *Notations for Software Design* Springer-Verlag, 1994, ISBN 3-540-19902-0.
10. R. Kazman, L. Bass, G. Abowd, M. Webb. *SAAM: A Method for Analyzing the Properties of Software Architectures*. In *Proceedings ICSE 16*, May 16-12,Sorrento, Italy, pages 81–90.
11. W.H. Wolf. *Hardware-Software Co-Design of Embedded Systems*. In *Proceedings of the IEEE*, Vol. 82, NO. 7, July 1994, pages 967-989.
12. *Vitrivius the ten books on architecture*. Dover Publications, New York, ISBN 486-20645-9.

13. M.E. Wiklund. *Usability in Practice: How companies develop User-Friendly Products*. Academic Press Professional, 1994, ISBN no: 0-12-751250-0
14. J. Nielsen. *Usability Engineering*. Academic Press, 1993.
15. V. Ambriola, R. Di Meglio, V. Gervasi, B. Mercurio. *Applying a Metric Framework to the Software Process: an Experiment*. In *Proceedings EWSPT'94*, LNCS 772, pages 207–226, 1994.
16. W.S. Humphrey. *Managing the Software Process*, Addison-Wesley, 1989, ISBN no: 0-201-18095-2.
17. F. Brooks. *The mythical Man-Month* Addison-Wesley, 1972, ISBN no: 0-201-00650-2.
18. G.P. Saxena, J.R. Engelsma. *Software Architectures: a Strategic Step towards 10X* Motorola, Strategic Software Technologies, Arlington Heights, IL 60004.
19. F. van der Linden and J. Müller. *Composing Product Families From Reusable Components*. In *Proceedings ECBS'95*, Tuscon, Arizona, IEEE to be published.
20. J. Müller. *Integrating Architectural Design Into The Development Process*. In *Proceedings ECBS'95*, Tuscon, Arizona, IEEE to be published.

Open Issues in the Design of PM Languages

C. Montangero

Dipartimento di Informatica, Università di Pisa, Italy

The Software Process community always played greater attention to the problems related to representation languages than other communities involved with modelling. This attitude follows from a cultural bias, which makes this community, as a part of the larger computing scientists community, aware of the importance of the tools, and especially of the linguistic ones.

This focus on languages was obviously induced by the goal of process enaction, which is peculiar and natural to the computing community, and also by the awareness that rigorous, if not formal, notations have an enormous leverage on the clarification of the modelling concepts and issues.

The papers in this section explore the current trends with PM languages. The paper by Conradi and Liu tackles the following problem: how can we best exploit the experience that, as computing scientists, we have in the design and implementation of executable languages, to develop effective process modelling *enactable* notations? The second paper of this session maintains that this is indeed a problem, challenging the recent ISPW9 position of Osterweil and Heimbigner, who instead favor a "federated" appraoch, where the emphasis is on the inter-operability of modules with different complementary functionalities.

Some interesting questions arise:

- Is the characterization introduced by Conradi and Liu adequate?
- How does it helps in assessing the past and, more importantly, the future linguistic proposals?

Another important facet of the problem is related to process evolution, that is universally recognised as essential requirement in Software Process Modelling. Some of the relevant questions, that should be considered in the discussion during the workshop, are the following:

- Which reflective characteristics make a single language well suited to deal with process evolution?
- Does the federated approach advocated by Osterweil and Heimbigner help, or rather it introduces stronger friction, hindering evolution?

Finally, the importance of formal reasoning about the properties of systems is becoming more and more appealing to the software producers. This raises a final question: how should a process modelling language be designed, in order to provide a good basis for the construction of a related logic system, to be used to reason about the models?

In Favour of a Coherent Process Coding Language

C. Montangero

Dipartimento di Informatica, Università di Pisa, Italy

1 Issue

The search for a universally useful process coding language should now be called off ... and substituted by ... a strategy of encoding an entire process as a distributed collection of modules, each addressing a restricted set of issues and facilitated by a special purpose language[1].

... we have to live with many sub-languages around a core process coding language. However, what counts is ... interoperability to handle heterogeneous and distributed process information ...[2]

2 Position

I will argue that the quest for such a language should not be abandoned in the long run, but only postponed, until the experience with the approach advocated above has identified coherent subsets of the domain that can be optimally integrated into a single whole.

3 Argument

3.1 Arguments Against Process Languages, and Proposed Alternatives

Let us first resume the arguments put forwards by Osterweil and Heimbigner (O&H in the sequel) in [4], against "a single formalism that should be expected to be effective in representing executable process specifications", i.e. a language that allows the efficient and correct implementation of efficient process-centred environments. O&H point out as a problem that "such a language needs to have so many, and so varied a set of, semantic features, as =8A likely to be impossibly large and complex", and that "were such a language available now, it is hard to imagine that programmers would very rapidly or readily become facile in its use". They propose to adopt instead "a strategy of encoding an entire process as a distributed collection of modules, each addressing a restricted set of issues, and facilitated by a special purpose language". The basic experimental

---

[1] Adapted from [4].

[2] Adapted from [1]: the authors use process modelling language, rather than coding, but the context makes clear that they are talking at the same level of [OH94].

support for the claim that such a strategy may be successful, is Arcadia, which has been implemented according to the suggested strategy. The strategic goal is to identify "a canonical set of process program execution support modules" to be implemented on different platforms. Quite similarly and independently, Conradi and Liu [1] (C&L in the sequel) are proposing that four approaches to PML design may be identified in principle: a fixed and large core (FLC), an extensible small core (ESC), a core language plus several compatible sub-languages (C&SCLs), and finally a core language and several incompatible sub-languages (C&SILs), and provide examples of each class. O&H's approach seems to fall in a fifth class, that we might characterise as that of the no core, several incompatible languages (SILs). We will also refer to this approach as the interoperability biased approach. C&L then proceed to provide evidence that now-a-days the design choice is no longer free, since part of the domain is "covered by existing or standardised languages and associated tools, like versioning or user interfacing". Therefore, they conclude that what is really needed is inter-operability, i.e. C&SILs. However, they also propose a segmented architecture, that breaks a PSEE in modules, interconnected among them and with the core model (i.e. the model represented in the assumed core language) via a CORBA bus. Although C&L's prose confers a special status to the core model, they add a picture that clearly shows that this is not the case, and that the process engine is just another module, as O&H maintain. In other words, they are also proposing a SILs approach. To conclude this section, we note that C&L's analysis implies that H&O's quest for a "canonical set of process program execution support modules" is only partially needed, since some of these modules are already there, like versioning or user interfacing, as mentioned above.

3.2 Arguments in Favour of Process Languages

The main reason to keep the design of PMLs in the research agenda of the Software Process community, is that it helps in focusing on at least two important subgoals, which stem from the very nature of a programming language. Indeed, we understand that, besides its concrete syntax, a process language is characterised, like any language, by i) its basic semantic domains, which define the primitive data types, the related operators, and the primitive actions; ii) its composition mechanisms, which define how to build complex processes from the basic ones; and, last but not least, iii) the associated logic system that allows to state the relevant properties of the process, to relate them to the process program and supports reasoning about this relation. The first natural subgoal that we consider is the characterisation of the basic semantics domains which are relevant to the software process, and their implementation on a number of platforms. We will see below, how the Arcadia [5] experience, as well as others, l= ike our own in Oikos [2], can be interpreted as a step in this direction= . The second natural subgoal that stems from the linguistic approach, is related to the composition mechanisms and to the logic system. The choice of the composition mechanisms is obviously critical for the implementation of efficient and correct processes: they must match the elements in the different basic domains well enough as to allow the efficient implementation of the process as a whole, but also to allow to analyse naturally in the logic system how they are used, to support correct implementations.

3.3 More Detailed Arguments

Let us consider the essential semantic features that O&H consider to be the root of the problems with the process languages to be[3]: "a rich object-oriented type system; powerful transaction management facilities; a rich resource specification mechanism; powerful concurrency and real-time features; powerful relation and relationship specification and management facilities; a flexible and powerful capability for specifying rules; very powerful, descriptive and flexible consistency management; planning features; pervasive dynamism". Essentially, O&H rely on the Arcadia experience to show that a number of those requirements for a process language are not so pervasive: they cite the components that deal with object management (PLEIADES), user interfaces (Chiron), measurements (Amadeus), process state (Process Wall). Experience with other systems also point in the same direction: for instance, we developed an Object Management System (OMS) in Oikos, to support the management of products, tools and workspaces. In these systems, there are indeed well identified components that respond to (well identified subsets of) a few requirements: for instance, PLEIADES in Arcadia, and OMS in Oikos provide the transaction management facilities advocated by the second requirement above. They also contribute to fulfil the first requirement, since the type system that is used for object management obviously adds to the overall type system, albeit as a well identified subsystem: the process programmer in Arcadia or Oikos, has some primitive concept of object to use in his endeavour, and some predefined constraints on how he can compose the objects and prescribe ways to access them. The interoperability biased approach interprets the situation as one in which there is a component of the enacting support that fulfils the requirement on transactions in the software process. The process modeller must deal explicitly with this component, and exploits whatever mechanism he has to guarantee the interoperability, e.g. RPC in Arcadia, and the low level EXPO communication protocol in Oikos. Finally, the interoperability biased approach foresees the independent design/development of the components per each process, with an eye to performance, with respect to the available technology. The linguistic approach takes a more abstract view of the same situation: one looks for subsets of the requirements that can be fulfilled by assuming a reasonably self-contained semantic sub domain, and implements it, adding as much syntactic sugar is needed to embed it in the general process coding language. The interesting point is that these domains are more complex that those of the standard programming languages, since they are biased towards the process. Therefore, the development of the corresponding logic sub theory becomes a non trivial task of its own, the more so when one realises that all

---

[3] We almost agree with this list, and the disagreements are not relevant here.

these sub theories must also be embedded in the general theory which is needed to reason on the process as a whole.

4 Conclusions

The conclusion I draw, with respect to the research agenda for the SP community, is that the emerging interest for the inter-operability of a set of independently developed modules should be cultivated, and will play an important tactical role, but should not be considered a strategic target, as should instead be that of identifying a core language (a few core languages, each) accompanied by several compatible sub-languages. The reason is that the experience with systems that interconnect independent modules will be an effective way to understand what is needed of each of them, as service providers, and what instead is there that should not, like hidden co-ordination policies and possibly incompatible redundancies. However, the quest for inter-operability alone reflects an optimistic laisser-faire attitude with respect to the evolution of PSEEs: let the individuals communicate and grow, and the community will also grow. I don't see any real evidence supporting this view. Also O&H show some doubts, when they state that "it will be interesting =8A to see whether interoperability difficulties appear and accelerates the drift [in the direction of increasingly different coding languages and platforms] continues". Rather, I see some analogy between interoperability and construction with natural materials: in both cases the engineer has to cope post-hoc with mismatches that can be instead avoided using carefully designed artificial bricks and sub-languages, respectively. As a final note, there is at least one candidate as a basis for the core language, namely the emerging co-ordination theory [3].

References

1. Conradi, R., Liu, C.: Process Modelling Languages: one or many? In these Proceedings.
2. Montangero, C, Ambriola, V.: Oikos: constructing process-centred SDEs. In A. Finkelstein, J. Kramer and B. Nuseibeh (eds) Software Process Modelling and Technology, Research Study Press (Wiley), Taunton, 1994, 131-151.
3. Malone, T.W., Crowston, K.: The interdisciplinary study of Co-ordination. ACM Computing Surveys **26** (1994) 87–119.
4. Osterweil, L., Heimbigner, D.: An Alternative to Software Process Languages. In C. Ghezzi, editor, Proc. ISPW9, Airlie, Va, 1994.
5. Kadia, R.: Issues Encountered in Building a Flexible SDE - Lessons from the Arcadia Project. In Proc. ACM Sigsoft SDE5, 169–180, 1992.

Process Modelling Languages: One or Many?

Reidar Conradi, Chunnian Liu

Norwegian Institute of Technology, Trondheim, Norway.
Beijing Polytechnic University, Beijing, P.R. China.

Abstract. The paper describes the different phases and subdomains of process modelling and their needs for conceptual and linguistic support, and in what forms. We group the relevant factors into three dimensions: meta-process phases, process elements, and the tool/user views. In the first dimension, we focus on enactable process models. For such models, we describe the design alternatives for a core process modelling language and a set of tailored sub-languages to cover special process elements. However, no detailed and functional comparison of possible modelling language are attempted.

Then we address interoperability between related sub-models and its implication to the language design. We also present a general architecture for a Process-Centered Software Engineering Environment, with a segmented repository of model servers.

Some concrete language realisations, mainly from the EPOS PSEE, are used throughout the presentation. We also give a realistic example of the design of an interoperable PSEE, and discuss how it can be improved using an extended EPOS.

The paper concludes that we have to live with many sub-languages around a core process modelling language. However, the underlying linguistic paradigm in this core language is not judged critical. What counts is use of standard support technologies, interoperability to handle heterogeneous and distributed process information, non-intrusive process support, end-user comprehension, and flexible support for evolution (meta-process).

1   Introduction

In the last decade, there has been much interest in the *software process* as a vehicle to improve software production. Many *Process-Centred Software Engineering Environments (PSEEs)* have been constructed and documented, although to a less extent tested in industrial settings. A few commercial PSEEs have also become available in the last years, e.g. Process Weaver from Cap [Fer93], and Process Wise Integrator (PWI) from ICL [Rob94].

The goal of this paper is to pragmatically discuss the design of the "ideal" *process modelling language* versus the functionality required. A special concern is how existing modelling concepts and linguistic constructs can be used and combined. Therefore, interoperability between different languages and related models and tools are emphasised. Some aspects of a general PSEE architecture are also discussed. However, the paper is *not* to give concrete technical solutions for particular problems in PML design. Rather, it takes the PML design as a whole, tackling the important issues – factors in process modelling vs. linguistic supports in PMLs, and the interoperability – in a general, perspective style.

Discussing the "right" modelling concepts and language(s) for a certain application domain is very ambitious, and the actual design of language can be hard to validate ("air castles"). We are faced with similar design decisions in Information Systems Engineering, Business Process Reengineering, Enterprise Modelling etc. – resulting in a plethora of modelling formalisms. See also [LSS94].

For instance, we may choose to have one large modelling language for the entire domain. Another alternative is to have one small and general core language, with a spectrum of specialised sub-languages to cover different subdomains (activities, artifacts etc.), to cover different goals (understandability, executability etc.), or to be oriented against different user roles (expert designers vs. normal users). We should also pay attention to standardisation and new platform technologies, cf. distributed architectures and workflow systems.

To obtain more solid assessments and comparisons of the modelling alternatives, we should have a well-defined set of domain subcategories, i.e. process elements. We will therefore focus on process modelling languages for rather "low-level" or enactable process models, not on the more high-level modelling languages to analyse and design more abstract process models.

There is yet little agreement on basic concepts, terminology, linguistic constructs, and tool architectures underlying such PSEEs. Some classification of concepts can be found in [CFFS92] [Lon93] [FH93] [CFF94], of languages in [CLJ91] [ABGM92], and of architectures in [BBFL94]. We will use the below definitions.

By *process* we understand a set of goal-oriented, partially ordered and interacting *activities*, and their associated *artifacts* and resources (humans, tools, time). Activities, artifacts, tools etc. are called *process elements*. We will limit ourselves to *software processes*, covering both development and maintenance of software. A process consists of a *production process*, a *meta-process* to regulate evolution of the whole process, and a *process support* (the PSEE) consisting of a process model and various process tools (Process Engine etc.). The production process is partly *external* to a computer, and is carried out by humans assisted by computerised production tools, e.g. CASE tools. The process support is normally *internal* to the computer, but may be supplemented by manual procedures. A PSEE has an explicit definition of the process, a *process model*. The model is expressed by one or several *Process Modelling Languages (PMLs)*, perhaps at various levels of abstraction or granularity (from template to enacting), and/or covering different parts of the process. It is usually stored in a PSEE repository, and consists of *model fragments*. The model can be interpreted to provide the user with process assistance of some sort (guidance, control, automation).

The paper will specially draw on examples from the EPOS PSEE [JC93] [C+94] and its process modelling language called SPELL. EPOS has been developed at NTH in Trondheim since 1989, first as part of a national project, later as a Ph.D. project. The emphasis is flexible and evolving process support during enactment for multiple software developers, using conventional development tools and working on files in checked-out workspaces. EPOS process models (task networks) are expressed in SPELL, an object-oriented, concurrent and reflective modelling language. The process models are stored in EPOSDB, a versioned software engineering repository. EPOSDB supports nested and cooperating transactions coupled to projects, and there is one Process Engine per transaction. The underlying platform is Sun workstations running Unix, and Prolog is the SPELL implementation language.

The paper is organised as follows. Chapter 1 is the above introduction. Chapter 2 discusses the process modelling subdomains, the PML design alternatives, and some solutions in EPOS and a few other PSEEs. Chapter 3 discusses the design of PMLs in an interoperability perspective, and presents a general PSEE architecture and an example to illustrate this. Chapter 4 contains a conclusion.

2 PMLs: Functionality versus Solutions

In this section, we will discuss the different design alternatives for PMLs, both generally and wrt. to the functionalities that must be covered. The EPOS PML(s) are used as an example, but with comparisons to PMLs in other PSEEs.

2.1 The PML Design Dilemma

Much research effort has been spent on designing the "right" PML. We can identify four approaches **L1–L4** for PML design, and all include a **core** PML, cf. [ACF94] :

- **L1: One fixed and large core PML**.
  Here, the core PML contains language primitives to express all relevant process elements. Although Occam's razor might be successfully used, a large but common PML will result.
  Typical examples of "hard" language primitives include constructs for activity triggering and concurrency.
- **L2: One extensible and smaller core PML**.
  Here, the core PML contains less primitives, rather a set of declarative constructs. Thus we can define tailored process models (often types and their instances), still within a common PML.
  Examples are "soft" descriptions of product structures, user roles, or tool interfaces.
- **L3: one core and several compatible sub-PMLs**.
  Here, many of the above process elements will be covered by separate and usually more *high-level* sub-PMLs. These may have well-defined interfaces to or be down-translatable to the core PML. We could also envisage an inverse translation from a more high-level core PML to alternative low-level sub-PMLs, e.g. to generate alternative implementations; cf. last comments in Section 2.2 and most of Section 3.

Examples are descriptions of check-out/in of workspaces, metrics collection, or transaction protocols.

- **L4: one core and several incompatible sub-PMLs**.
  Here, such sub-PMLs will be separate languages, but wholly independent of and often orthogonal to the core PML.
  Example are descriptions of versioning or user interface paradigms.

The chosen strategy for PML design will influence the size and complexity of the PML and the resulting PSEE. A good language design assumes that we understand the domain, so that we can make sensible decisions on which process elements should be covered where and how. If our knowledge is poor or immature, we might initially experiment with a small core PML and many sub-PMLs. After collecting experiences, we are in a better position to decide – both strategically and technically – how the mutual sub-PML interfaces should be, which sub-PMLs could be reconciled, or which ones could be merged into the core PML. There is clear analogy with conceptual modelling of software systems: In newer approaches (see e.g. [vV92]), entity-relationship or object-oriented data models have been effectively combined with data flow diagrams or Petri-nets to describe the static and dynamic parts of the domain, respectively. Cf. also *federated databases*, trying to unify different data models, schemas and instances from possibly heterogeneous subdatabases.

However, sometimes we do not have a free design choice, since parts of the domain already are covered by existing or standardised languages and associated tools. The challenge is how the core PML and its PSEE shall or can *interoperate* with non-core languages and tools, e.g. for user interfaces, configuration management, or project management. This means that we should think strategically to prepare for co-existence of possibly inhomogeneous and partial process models. We may even stick to a small core PML to reduce labour and risk.

In other words, there are many factors to consider, when specifying and assessing the functionality of a PML to describe a given or desired process. In Figure 1 we group these factors along three main dimensions: meta-process phases (design, implementation etc.) process elements (activities, artifacts, ...), and user/tool interaction. In the "process element" dimension, there is a list of core and non-core process elements and their PML design approaches (**L1-L4**), which can be regarded as a summary of Section 2.3 and Section 2.4. In addition comes general PML design criteria, such as understandability and modularisation (Section 2.5). For all these dimensions, the EPOS PSEE and some other PSEEs are commented. However, the emphasis is on their positioning of the core PML vs. the sub-PMLs, not on comparing specific language paradigms.

The next Section 3 contains a more general discussion on PMLs in the perspective of interoperability, presenting a general PSEE architecture and an example of this.

2.2 PMLs and Meta-process Phases

The PML could be used to support the following meta-process phases:

Meta–Process Phases

Process Elements

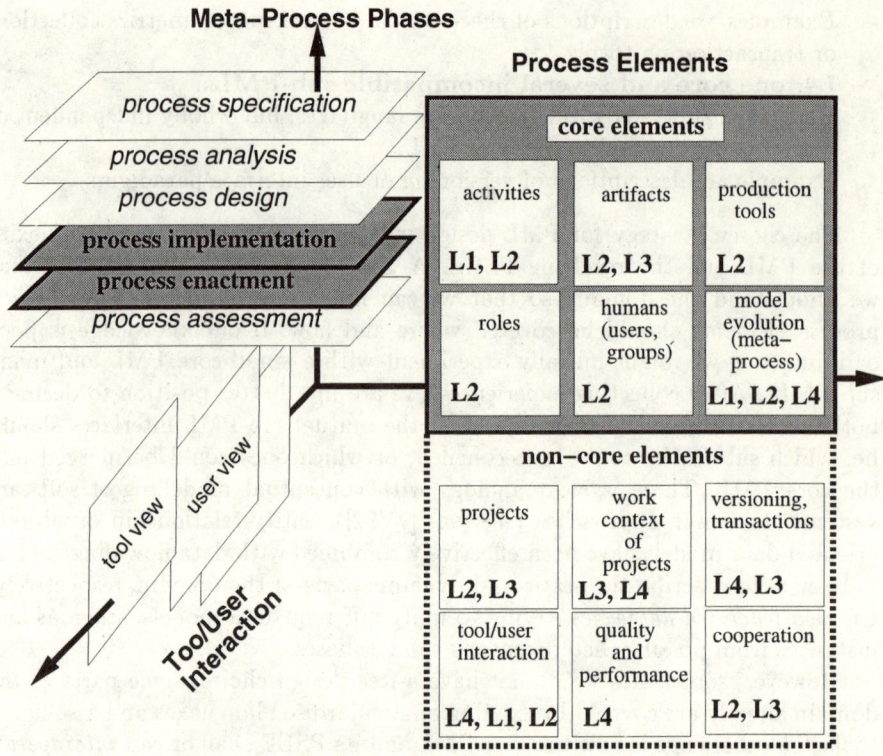

Fig. 1. Three Dimensions of PML Design

1. **Process elicitation and requirement specifications**:
   Here, we must assist human understanding and negotiation of the perceived *as-is* process. We will need overall (conceptual) modelling, often stating business rules, coarse-grained work flow, and general work responsibilities. Intuitive and often graphical notations may be important if the audience is company decision makers. The PML of this phase will resemble languages used for Information Systems and Enterprise Modelling.

2. **Process Analysis**:
   Here, the PML must be sufficiently formal to be used for reasoning or simulation (dry runs). Changed needs from markets and new inputs from technology providers may enter in this phase. Decisions on changes to define a *to-be* process will typically take place here, so quality and performance matters must be dealt with here.

3. **Process Design**:
   Here, the PML must be able to express a more detailed process architecture and to incorporate more project-specific information, e.g. number of work-packages/subprojects, over-all planning (dependencies, timing), development technology (OO techniques), quality model etc.

4. **Process Implementation**:
   Here, the PML must allow specification of sufficient low-level details to achieve an enactable process model. This model must possibly be translated and otherwise prepared for execution. This may imply that parts of the process model is implemented and installed outside the PSEE, e.g. as tool configuration tables in a broadcast message server (BMS[1]) or as triggers/monitors in the production workspace.

5. **Process Enactment**:
   First, we start a Process Engine to achieve an enacting process model, residing in the PSEE repository. This Process Engine interacts with production tools and with human agents (wet runs). This interaction occurs, respectively, through a tool interface (e.g. a BMS) and a user interface (e.g. through an agenda).

6. **Process Assessment: Quality and Performance issues**:
   This covers anything from trivial follow-up of tool activations to collecting prepared measurements. All such data can be given as feedback to previous phases, either to guide the process or possibly to evolve the process model and its process. A Quality and Performance model must regulate all this, and the production tools must be properly instrumented for this purpose.

As shown, the different process lifecycle phases (the meta-process) exhibit different goals, contexts, and clients. It is doubtful that one PML can be used for all this, although we can use flow-based notations in most phases. Specialised and more high-level specification languages, with successive and not always automatic transformations towards more low-level enactable languages, is envisaged (the **L3** approach) – as for the normal software lifecycle. However, what is normally considered conceptual models in Information Engineering Systems are often being interpreted in PSEEs to support real enactment. For instance, specification languages like Petri-nets are used for such enactment.

In the following, we will only deal with PMLs for Process Implementation and Process Enactment. These are the only meta-process phases that we understand to some extent, but they also have the most detailed and heterogeneous information. Most PMLs and PSEEs, including EPOS, have emphasis on these phases.

2.3 PMLs and Process Elements to be Covered

A real-world software (production) process can be very large and have a variety of process elements. These must be modelled by a PML and supported by the associated PSEE. The PML should also describe the interface between the process support and the production process, covering tool activations, capturing of performance data, workspace set-up etc..

The next subsections will present a short-list of process elements (process subdomains), thus constituting a small **taxonomy** for such. These elements

---

[1] BMS is only used as an abbreviation, not as a potential product name.

must be covered by a core PML, or by a collection of sub-PMLs. Such a division implies a core process model and many sub-models. As will be shown in section 3, such partial models could reside on separate model repositories or -servers.

In our opinion, the **core PML** must support the following process elements, E1–E6: **activities, artifacts, roles, humans, production tools**, and **support for evolution (meta-process)**. The remaining process elements, E7–E13, should be covered by sub-PMLs outside the core PML, but with clear interfaces both ways.

E1. Activities (core)

Concurrent and cooperating activities are the heart of any process. They can be at almost any granularity level, and are usually associated to roles that can be filled by certain users and/or tools. Artifacts constitute the operands (inputs/outputs) of activities, so the core PML must contain a Data Manipulation Language (DML) to access such artifacts.

EPOS offers a predefined root task type (**L1**), that can be subtyped (**L2**) to express different model instances of tasks (in short: tasks). A task type is a normal object type but with special semantics attached to type-level attributes, such as PRE, POST, CODE, FORMALS and DECOMPOSITION (**L1**). External activities are modelled by cooperating and parallel transactions, containing internal tasks as co-routines. A task is runnable if its PRE-condition evaluates to **True**. The tasks stand in a chained and decomposed network, and is connected to artifacts, tools and users. A Planner tool helps in building such networks.

In SPADE [BFG93], an extended Petri-net formalism is used (**L1**) and with clustering of natural net units (**L2**). In Marvel [BK92], activities are described by a dual model: production rules resembling types (**L1**) and a separate network formalism (**L3**) to connect the rules and to attach the artifacts.

E2. Artifacts (core)

The artifacts describe the product in question. In a reflective system, all process model fragments can be considered artifacts.

The product model will usually contain a basic data model, a product schema, and instances of the latter. At least product composition and dependencies must be described. We generally advocate an object-oriented paradigm (cf. CORBA [Obj92] and ODMG [Cat94]), in spite of lack of final standardisation.

On modelling of artifacts, we face a strategic choice. One alternative is to model a product as mere placeholders (like in Process Weaver, **L3**), in order to be independent of the actual product workspace. Another alternative is to model the product rather extensively as in EPOS, Adele and Marvel. The latter allows some planning of project breakdown and suggestion of cooperation/propagation protocols, all essential for process modelling. (Note, that Process Weaver is being coupled to Adele in the PERFECT ESPRIT project.)

EPOSDB has a structurally object-oriented data model, being extended by the SPELL PML to full object-orientation. SPELL is used to define a system-defined product schema (**L2**), supporting hierarchic families with own interfaces and bodies. Some utility tools utilise this product model and an external product description language is defined (a **L3** sub-PML), see below on workspace organisation in Section 2.3.

E3. Roles (core)

A **role** defines responsibilities and rights for users that play that role.

In EPOS, a role is modelled as a separate model fragment, specifying certain access rights. Roles are then indicated by type-level attributes in task types (**L2**), and a task instance can only be connected to a user that can fill this role. No other semantics is utilised for roles, e.g. as in PWI.

E4. Humans: Users and Groups (core)

A human **user** or process agent can fill a set of roles. He can also be a member of several **groups**, possibly nested, representing either project teams or line organisations.

In EPOS, users are modelled as `Person` instances, having a dynamic role-set (**L2**). This is checked against the role specification of the task, currently being performed by the user. No such groups are p.t. implemented.

E5. Production Tools (core)

The tool model must specify how such tools can be accessed and controlled. We must distinguish between batch and interactive tools, and be able to handle call-backs from both.

In EPOS, a production tool is modelled as a low-level task type, specifying I/O, command line formats, options etc. (**L2**). A tool is started by sending a detailed message from the task's CODE part. The response is returned through an extra task input, and causes the internal task to be reactivated.

SPADE offers more flexibility to describe tool behaviour as a cluster of Petri-net transition nodes (still **L2**).

E6. Support for Evolution or Meta-process (core)

Due to the human-oriented nature of the software process, we have an inherent cause for evolution during process enactment. This means that most previous lifecycle phases must be repeatable "on-the-fly". Thus the core PML must offer support for evolution of at least the process model, both technically (e.g. by reflection or interpretation) and conceptually (by a defined meta-model). Evolving an *enacting* model is known as "pulling the rug" from telecommunication switches, relying on dynamic linking and special hardware.

The available PML mechanisms to support model evolution differ, e.g. by predefined facilities (delayed subtask expansion in MELMAC [GJ92], **L1**), by a fully reflective PML (as in EPOS, SPADE, and PWI – **L2**), by another language (as in Marvel, **L4**), or ad-hoc (Process Weaver, no support). Note, that parts of the process model may exist in many different (translated) forms in a PSEE and its environment, cf. Section 2.3. Thus, to have reflection or interpretation in the core PML, as implemented by some PSEEs, is grossly insufficient.

However, none of the PSEEs deal explicitly with evolution of the real-world production process, also being an enacting and concurrent body.

E7. Projects (non-core)

A project contains a variety of domain-specific information. A project model thus might include: a workplan with sub-activities/-projects, responsible for activities, overall goals and inputs/outputs, available resources, time estimates, work records (time sheets, logs etc.), quality model (Section 2.3), cooperation patterns between projects, and connection to workspace transactions and versioning (Section 2.3). Such project information is revised almost daily (i.e. it is highly "versioned"!), both to record ongoing work and to make adjustments based on this.

Evidently, some of this information (**L3**) overlaps with or is translatable down to the core process model (**L2**).

In EPOS, a project model fragment is of the `project` task subtype (**L2**). It is connected 1:1 to a PSEE repository transaction (**L3**).

E8. Work Context of a Project (non-core)

This includes a production workspace and the available production tools.

A **production workspace** (e.g. files or a CASE-tool repository) is the external representation of a configuration, which again is a requested part of the total, versioned product. A mapping and check-out/check-in between the internal PSEE repository and the external workspace may be needed. The **production tools** have possibly been instrumented with interfaces to the process tools.

An EPOS production workspace is a set of files checked out from the EPOSDB repository through a Workspace Manager, which has a special check-out tool and an EPIT production description language (**L3**). The configuration specification and the mapping from the repository to the workspace is described in special languages (**L3** and **L4**), being a part of the previous project model (Section 2.3). In case of overlapping workspaces, the Cooperation Manager must be activated (Section 2.3). Production tools (e.g. compilers, editors) can then work on these files, guided by the PSEE. Before transaction commit, the modified files must be brought back through a similar check-in tool.

E9. Product Quality and Process Performance (non-core)

A **product quality model** includes operational goals of product quality and associated metrics, e.g. review and test status [IS94]. The **process performance model** for process quality expresses compliance to the stated process model, e.g. wrt. deadlines, budget, and user roles.

EPOS has none of this now, but this is being worked upon in two PhD theses.

E10. Versioning and Transactions (non-core)

Versioning at least of the production workspace is needed, and likewise with some support for long transactions. Whether such support should be extended to also encompass the PSEE repository is an open issue.

The chosen **versioning model** should allow transparent versioning during a transaction. It should be orthogonal to the process model, so that the same versioning can apply uniformly to all process model fragments. Note, that there are parts of the process support, that cannot easily be subjected to formal versioning and transactions. We can mention configuration tables in BMSes, "shell-envelopes" around production tools, or extra triggers in production workspaces.

EPOS rely on Change-Oriented Versioning [L+89] [Mun93], giving uniform versioning. Process modelling and versioning work well together (**L4**). On the one hand, a particular version of a partial product can be checked out into the workspace before the process starts, using EPOSDB facilities and the added check-out tool (Section 2.3). On the other hand, the EPOS process model is itself a first-class artifact, and can be stored, versioned and reused by the same versioning model.

In Adele2 [BEM93], the product model instances (product descriptions) are supported by one versioning model. For the product schema and for the activity/tool model, other facilities are used to achieve evolution, e.g. special delegation paths and roles (meaning versioned types).

The **transaction model** should be nested and allow pre-commit cooperation. Preferably, the PSEE repository should contain an instrumentable transaction model, with "hooks" into the process model in case of certain events. EPOS has extensive support for cooperating transactions, where high-level cooperation protocols (**L3**) are translated down to e.g. operational triggers and task networks. Similar has been demonstrated by Adele. EPOS also provides some planning tools for transaction (or project) breakdown, scheduling, and cooperation [CHL94].

E11. Cooperation (non-core)

We have two basic modes of cooperation: *sequential,* e.g. by normal work or review chains, or *parallel,* e.g. upon workspace overlap.

For sequential cooperation, the normal EPOS task networks are used, also across transactions. For parallel cooperation, EPOS has a Cooperation Manager that helps in setting up high-level cooperation policies (**L3**) between projects/transactions. These are then translated into more low-level support (triggers, task networks, inter-transaction propagation – i.e. **L2**), using the extended transaction concept of EPOSDB (Section 2.3).

Using roles to express interaction diagrams has been investigated by e.g. PWI, but not by EPOS.

2.4 PMLs and Tool/User Interaction Paradigm (non-core)

This deals partly with how and to what degree the process support intervenes with the user's normal way of working, called the *tool view* (or PSEE tool coupling). This is only relevant in the last lifecycle phases.

It also deals with how parts of the process model should appear through a *user view*, both conceptually and graphically. This applies in all lifecycle phases, but especially in the last ones.

E12. Tool View (non-core)

The tool view deals with how much the process support "perturbates" the production process. We can identify at least two different tool views or work modes [FG94]:

1. In most existing PSEEs, the user has a *task-oriented view* of the process: The user is directed by the PSEE. That is, most tool activations are strictly controlled and invoked through a central process support interface. That means that all relevant production tools must be enveloped.
2. More preferably, the user needs a *goal-oriented view*: After setting a goal, he can move freely in the process space to achieve the goal. The process support will listen to events in the production process, and give guidance based on this. Again, the production tools must be (invisibly) enveloped, or all relevant accesses to the production workspace must be trapped to inform the PSEE.

The tool view often gets embedded in the PSEE, both conceptually and technically, although this is hardly ever explicitly expressed in the PML (thus **L1**). Is might have been possible for the user to select between the two above alternatives (assistance vs. control), and even to do this dynamically (!).

EPOS allows either manual or automatic activities to be expressed in SPELL. The former are approved through an agenda, that contains runnable activities with **True** PRE-conditions (partly **L2**).

The Marvel Process Engine only approves user actions (**L1**), but cannot itself initiate actions behind the user's back, not even recompilations.

E13. User View (non-core)

The user view or user interface is a crucial, but often neglected part of many PSEEs, including EPOS. This fact is amplified by the delicate balance between "control" and "assistance", cf. the previous subsection. However, the problem of displaying complex and heterogeneous information to users with different levels of competence and goals, are shared by most computerised systems [Mye89].

The general paradigm of user interfaces is that we should split *how is works* (internal model, e.g. in C or Prolog) from *how to do it* (external view) – i.e. **L3** or even **L4**. Some aspects to consider are: uniformity of presentation and interaction, easy comprehension of presented information, and flexible choice of presentation formats (filters, viewers). For instance, the external view could be a task network, while the internal activity model is rule-based – or inversely! Cf. also the ECMA roaster model [ECM91], where the user interface is made separate. This separation has a tactic consequence: first we should define by fast prototyping the interface on the user's premises and based on familiar concepts according to his role. Then we may consider to implement this interface by some suitable technology.

Much research has been devoted to expressing multiple and possibly incompatible model views to the user. The views can be conceptual (using ViewPoints [NKF93]) and/or graphical (using Smalltalk's Model-View-Controller [GR83]). For the latter, [Fol83] recommends showing simply the entities and their relations and operations.

EPOS provides the user with an activity network and an agenda. The network is a simple print-out of the internal data of the process engine. A viewing filter between the engine and the user is needed, providing alternate views according to user roles.

2.5 General PML Demands (core)

Analysability: A PML, as most modelling languages, should be sufficiently formal to allow precise modelling, analysis and simulation. Most PMLs have this property, including the EPOS SPELL.

Understandability: A PML should be user-oriented and easy to comprehend. This may assume informality to some extent, or separate presentation languages (**L3**). However, most PMLs are too low-level and technical, including the EPOS SPELL.

Modularisation: It should offer modularisation features to support grouping and protection, and sub-model constructs to facilitate reuse on a rather high level, e.g. as sub-schemas (**L2**). This point is further elaborated in the next section on interoperability (**L3** and **L4**). EPOS allows models to be organised in task hierarchies, e.g. according to projects.

2.6 A Short Summary

A short summary of the core and non-core process elements and their sub-PML design approaches discussed in this Section can be found in Figure 1, along the "Process Elements" dimension. Note, that many PSEEs do not conform to the recommended core PML design, as **L1** and **L2** constructs are used to express non-core process elements.

In general, the question *"what is core and what is non-core?"* is certainly a disputable one. We present our views here based on our research and experiences in SPT area, and hope that our views will trigger interesting discussions in the SPT research community.

In the next section, we will propose a PSEE Architecture, based on our views of interoperability. What we said above for the core/non-core distinction is also true for the architecture – we hope that it will trigger similar discussions in the SPT community.

3 A General Perspective on PML Design: Interoperability

As mentioned, we may need a core PML and a set of sub-PMLs to cover the variety of process elements, or to interface to existing and domain-specific models, their modelling tools and model repositories. In most PSEEs, all the (sub)models are stored in a common repository, so it is worth to consider segmentation also of this repository.

Figure 2 shows a general PSEE architecture with a segmented repository, or rather a collection of **model servers**. A relationship model server could be added to store inter-model references. The segmentation could be conceptual with all sub-models still residing in a single repository, or physical with some sub-models in separate repositories. This means that the associated PSEE tools become tools in their own right, not just local procedures in a common process tool. Note that the ECMA CASE environment architecture is applied throughout, with separate database-, tool- and user interfaces.

In our opinion, most of these sub-models should, at least representationally, be defined by a *(structurally) object-oriented* data model. The models should be accessible to the PSEE tools through a standard object-oriented database interface, as defined by CORBA or ODMG.

This section considers interoperability between the different sub-models and the PSEE tools, and between these and the production workspace/tools. The emphasis is on the implication to PML design.

3.1 Interoperability Between the Core Model and the Other Sub-models

Core vs. Project Model

As mentioned, interoperability between the core model and the project sub-model can be achieved by the **L3** strategy of PML design. This is done in EPOS and Process WEAVER.

However, often an existing project management tool must be incorporated into the PSEE, and this tool is often instrumented to the client organisation. In this case, EPOS must give up most of its project sub-model, and try to integrate (still **L3** level) with the former. In Section 3.2 we will present a real-world example of this.

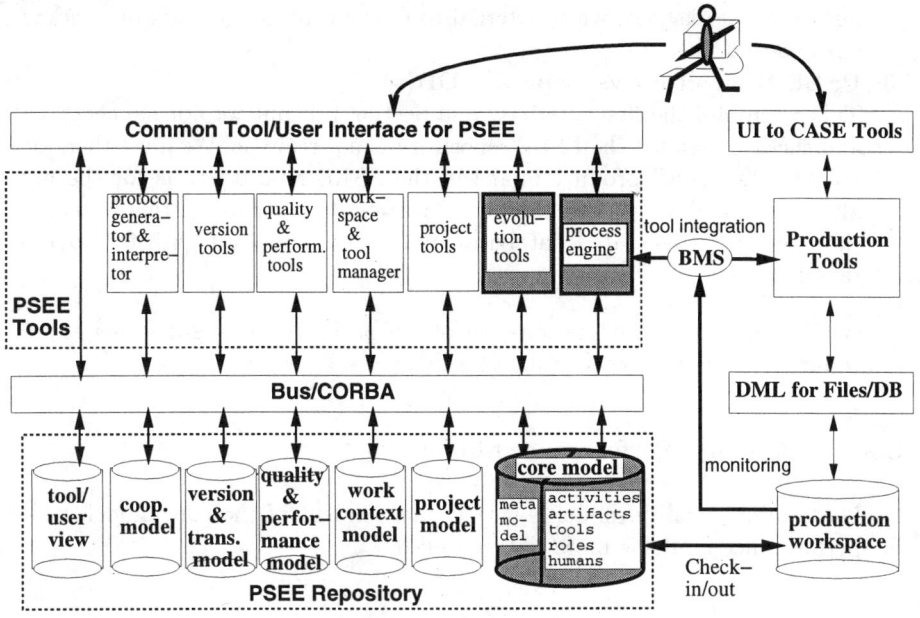

Fig. 2. PSEE Architecture with Segmented Repository.

Core vs. Work Context Model

The PSEE repository is almost always a software engineering database (DB), preferably object-oriented. We can identify three possible combinations of the PSEE repository and the production workspace:

1. **PSEE Repository vs. File System**:
   As mentioned before, one alternative is to do check-out and check-in of files from the PSEE repository as part of a transaction, and to ensure some level of mutual consistency during this transaction. All this require specifications in and translation from a **L3** PML.
   Another alternative is to rely on a *virtual* file system, where all file accesses are trapped and mapped directly onto the PSEE directory, cf. PCTE [BGMT88] or ClearCase [Leb94]. Transparent versioning is also easy to arrange with such a solution.

2. **PSEE Repository vs. own Sub-Repositories**:
   The "cleanest" combination is that the production workspace is a sub-database of the core PSEE repository. However, CORBA and ODMG technologies are immature, hence few production tools have adapted to these so far.
   Of PSEEs, Adele and EPOS have made the workspace *conceptually* a sub-database of a single repository. Most DBMS commands can then be carried

out in this workspace, while external tools can continue to work on checked-out external files.

3. **PSEE Repository vs. separate DB(s)**:
   This resembles the first alternative in Section 3.1, but we normal check-out and check-in against the PSEE repository is not realistic. We must therefore create a "shadow" product model in the PSEE repository, as for the first alternative and as in the example in Section 3.2. Generally, this brings up the question of federated databases, but this is not dealt with for obvious reasons.

In all these three combinations, an object-oriented paradigm seems appropriate for the PSEE-internal artifact model, i.e. **L3**.

Core vs. Quality/Performance Model

The product quality model is largely independent of the core model, while the performance model is more tightly linked.

Core vs. Versioning/Transaction Model

Our EPOS experience advises uniform and transparent versioning to the entire PSEE repository in a transaction context. However, most standard DBMSes or file systems do not offer such. Thus, the next-best solution is special versioning of certain parts of the repository or only for a file-based production workspace.

Versioning/transaction description languages (**L3/L4**) and tools are therefore pervasive technologies. They are linked to the core process model and also to the project and cooperation model.

Core vs. Cooperation Model

Cooperation is rather enhanced by interoperability, but the basic cooperation model should not be affected. An increasing number of groupware or workflow systems are also becoming available, with own languages (**L3/l4**) and paradigms. To make these accessible in the context of a PSEE will be a challenge.

Core vs. Tool View Model

The prevailing tool interaction technology is to use point-to-point (e.g. CORBA) or broadcast message servers (e.g. FIELD) to let tools cooperate in a flexible way. By proper enveloping, tools need not be aware of that they are "coordinated", nor do the PSEE's tool model or Process Engine be bothered by low-level details here. This means that tool descriptions are partly represented in the PSEE tool model and partly in BMS-internal configuration tables (**L3**).

Interactive tools like Emacs are difficult to handle, partly because they can do almost anything. It is, however, possible to trap and instrument their file accesses, or to rely on more fine-grained tool-tool interfaces.

Core model vs. User View Model

As mentioned, the User View or User Interface (UI) model is a critical part for a PSEE. The underlying linguistic paradigm of a PML is judged not crucial for a good UI (**L4**).

There are large savings by using existing technologies for user interfaces, as exemplified by X and Motif, UIMSes, graphical browsers and editors, and report generators. Thus, the interface between the User Interface and the PSEE tools should be standardised (**L3?**). Then we can more easily use standardised user interface technologies, and plug new process tools to the PSEE user interface.

3.2 An Example of a Segmented PSEE

The SYSDECO software house in Oslo has developed and is selling a 4G-tool called Systemator, running on Unix workstations. Using this 4G-tool, applications can be written in a special Sysdul language. From Sysdul "programs", Systemator can generate complete user applications against a given, commercial database.

Sysdeco is also developing customer applications in Systemator on behalf of customers, say in Project X. They use no computerised project management system to support such projects. A MicroSoft/Project tool on PCs is in use only by Project Managers, first to make a coarse plan – with activities, schedules and personnel – and then to manually record ongoing activity status (time sheets etc.). Most project data are recorded on paper. A quality model is defined, but again followed up manually.

Below, we will recapitulate the projected design of a PSEE demonstrator to offer more automatic support for such development projects. The work is done at NTH by a student project group, lead by Geir Magne Høydalsvik [Hø94]. For this demonstrator, Sysdeco insists on using only commercially available tools and databases for this, including their own Systemator tool. Thus, there is initially no room for EPOS.

Figure 3 shows the projected and segmented PSEE for this project support. The part drawn in slender lines represents the PSEE, while the part drawn in the bold or dotted lines represents a possible generalisation using an extended EPOS. Thus, the projected PSEE has the following components, where the actual PML being used is indicated as $L_{...}$:

1. **Systemator 4G-tool**, utilising L_{SYSDUL} and generating applications against a **product database** on top of the Ingres DBMS. This is the same as before, but triggers must be inserted in the database (expressed in L_{INGRES_DML}) to send messages to Process WEAVER upon certain events.

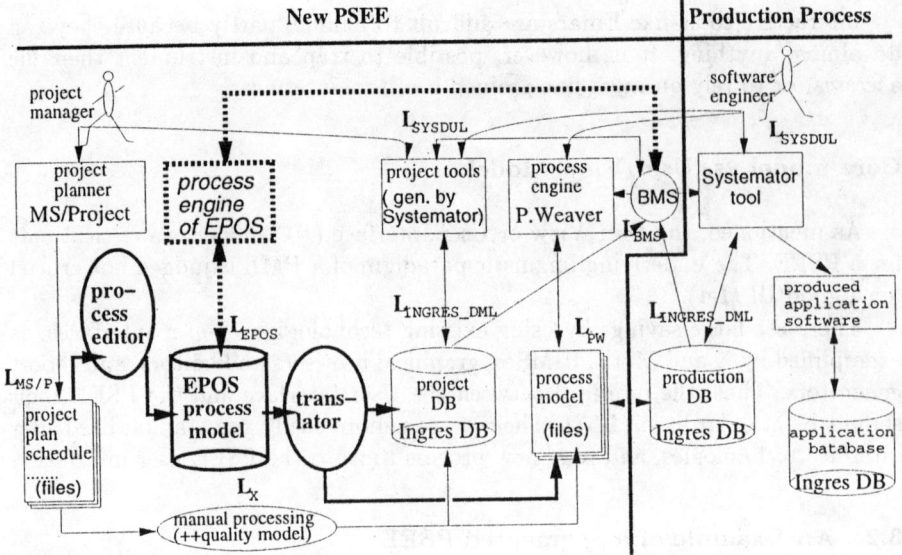

Fig. 3. Design of a PSEE for Project Support

2. **A MicroSoft/Project** tool, initially used by a Project Manager to produce a coarse project plan, resembling a Gannt diagram. This initial plan is printed out in some intermediate, legible format ($L_{MS/P}$). This plan is then manually extended into a full project model (in "L_X"), including an enhanced quality model. Probably, "L_X" will not be an entirely formal language.

3. **Project tools**, using a project database also implemented by Ingres. These tools will initially be generated by the same Systemator tool (and expressed in L_{SYSDUL}). The project database must similarly be instrumented by triggers (in L_{INGRES_DML}).

   These tools will first be used to install the full project/quality model. Later, the developers themselves will be filling in quality and performance data during normal development. Likewise, product data are being fetched from the product database, possibly through triggers.

4. **Process WEAVER**, being the Process Engine and using files as its process model repository (in L_{PW}). Process WEAVER will communicate both with the project tools and their project database, and with Systemator and its product database. Its process model will be manually made, based on the full project/quality model.

5. **A Broadcast Message Server** (configured by L_{BMS}), to link Process WEAVER with the other tools and databases.

6. Various graphical presentation languages (L_{UI}), not elaborated further.

All of these languages are classified as **L3** languages wrt. a core PML, perhaps with exception of L_{PW} being an enactable PML in its own right.

Our first observation is that this PSEE design fits very well with the general PSEE architecture from Figure 2. Let us then consider how more advanced process modelling techniques such as EPOS can be applied to improve (or enrich) the projected PSEE.

We can rather build a complete process model in a possibly extended SPELL (L_{EPOS}), based on the extended process/quality model. This combined and more high-level process model can then be translated into the above sub-models in their actual sub-PMLs, thus replacing the current manual processing. Indeed, we need a powerful **PSEE meta-tool** or CASE tool for process modelling. This should be able to build, translate, install and maintain a set of distributed and heterogeneous process models, used by partly commercial and standardised process tools! Note a possible bootstrap (meta-process), as the goal of the application project could be to make a new version of the project tools!

We can also use our EPOS process engine (in dotted lines of the figure) to instrument the production process. Of course, in this case we have to achieve interoperability with the existing project model.

4 Conclusion

Until recently, much research in process modelling has been focussed on different linguistic paradigms for the core PML, in order to find *the* correct one. This paper addresses the "one or many PMLs" question from a broader point of view.

To recapitulate: there is a big variety of process phases and elements to be covered, although we focus only on the Imnplementation and Enactment phases. We must also interface towards actual production tools/workspaces. Different user roles have different needs, also wrt. work modes and user interfaces.

There are many technical arguments behind choosing *one core PML* (**L1/L2** approach) and *a set of sub-PMLs* (**L3/L4** approach). However, the decisive factor in choosing the "federated" or "interoperable" approach, is that we *have to* adapt to a myriad of relevant through alien languages, tools, and databases. All of these must somehow be incorporated into or interfaced against the PSEE. That is, the PSEE developers simply do not control the PML design space. Thus, interoperability against standard or existing subsystems is an *absolute must*, specially since process support should be an add-on to existing computerised production tools, not a hindrance.

We even claim that the choice of the underlying linguistic paradigm for such a core PML (which must cover the short-list in Section 2.3) is not so important. What really counts is:

- **Standardisation**:
  Use of standard support technologies: Unix/MS-DOS, C++, CORBA or OBMG, X/Motif, new workflow systems, etc. etc.. Reuse is a keyword here.

- **Interoperability**:
Making PSEE components interact smoothly with other process and production models and tools. Modularisation, open systems and above standardisation are keywords here.
- **Tool view (PSEE coupling)**:
The PSEE should "perturbate" a software production environment in a minimal way. The main goal of the PSEE should be to give flexible enactment support at the appropriate level, e.g., control, automation, guidance, reasoning, explaining etc..
- **User view (user interface)**:
The process agent should be presented a comprehensible view of his current work context, with proper connections to the co-workers' activities. Been given this, the agent can (more) intelligently execute and relate to his own role in the overall process.
- **Easy user-level evolution of the process model**:
Again, the goal is to provide an understandable view of the model, so that this can be changed by the process agents themselves, if and when needed.

Acknowledgements

Thanks go to colleagues in the PROMOTER project, and to the local teams behind EPOS. Liu's work is partly supported by the national Natural Science Foundation of China (NSFC).

References

[ABGM92] P. Armenise, S. Bandinelli, C. Ghezzi, and A. Morzenti. Software Process Representation Languages: Survey and Assessment. In *Proc. 4th IEEE International Conference on Software Engineering and Knowledge Engineering, Capri, Italy, June 17-19. 31 pages*, June 1992.

[ACF94] Vincenzo Ambriola, Reidar Conradi, and Alfonso Fuggetta. Experiences and Issues in Building and Using Process-centered Software Engineering Environments, December 1994. Internal draft paper V3.0, Univ. Pisa / NTH in Trondheim / Politecnico di Milano, 28 p.

[BBFL94] Sergio Bandinelli, Marco Braga, Alfonso Fuggetta, and Luigi Lavazza. The Architecture of the SPADE Process-Centered SEE. In *[War94]*, pages 15–30, 1994.

[BEM93] Noureddine Belkhatir, Jacky Estublier, and Walcelio Melo. Software Process Model and Work Space Control in the Adele System. In *[Ost93]*, pages 2–11, 1993.

[BFG93] Sergio Bandinelli, Alfonso Fuggetta, and Carlo Ghezzi. Software Process Model Evolution in the SPADE Environment. *IEEE Trans. on Software Engineering*, pages 1128–1144, December 1993. (special issue on Process Model Evolution).

[BGMT88] Gerard Boudier, Ferdinando Gallo, Regis Minot, and Ian Thomas. An Overview of PCTE and PCTE+. In *Proc. of the ACM SIGSOFT/SIGPLAN Software Engineering Symposium on Practical Soft-*

ware Development Environments, Boston, Massachusetts, November 28–30, pages 248–257, 1988.

[BK92]   Naser S. Barghouti and Gail E. Kaiser. Scaling Up Rule-Based Development Environments. *International Journal on Software Engineering and Knowledge Engineering, World Scientific,* 2(1):59–78, March 1992.

[C+94]   Reidar Conradi et al. EPOS: Object-Oriented and Cooperative Process Modelling. In *[FKN94],* pages 33–70, 1994. Also as EPOS TR 198, NTH, 31 Oct. 1993, Trondheim.

[Cat94]   Rick G. G. Catell. *Object Data Management: Object-Oriented and Extended Relational Database Systems.* Addison-Wesley, 1994.

[CFF94]   Reidar Conradi, Christer Fernström, and Alfonso Fuggetta. Concepts for Evolving Software Processes. In *[FKN94],* pages 9–32, 1994. Also as EPOS TR 187, NTH, 9 Nov. 1992, 26 p., Trondheim.

[CFFS92]   Reidar Conradi, Christer Fernström, Alfonso Fuggetta, and Robert Snowdon. Towards a Reference Framework for Process Concepts. In *[Der92],* pages 3–17, 1992.

[CHL94]   Reidar Conradi, Marianne Hagaseth, and Chunnian Liu. Planning Support for Cooperating Transactions in EPOS. In *Proc. CAISE'94, Utrecht,* pages 2–13, June 1994.

[CLJ91]   Reidar Conradi, Chunnian Liu, and M. Letizia Jaccheri. Process Modeling Paradigms. In *Proc. 7th International Software Process Workshop – ISPW'7, Yountville (Napa Valley), CA, USA, 16–18 Oct. 1991, IEEE–CS Press,* pages 51–53, 1991.

[Der92]   Jean-Claude Derniame, editor. *Proc. Second European Workshop on Software Process Technology (EWSPT'92), Trondheim, Norway. 253 p.* Springer Verlag LNCS 635, September 1992.

[ECM91]   ECMA. A Reference Model for Frameworks of Computer Assisted Software Engineering Environments. Technical report, European Computer Manufactoring Association, 1991. ECMA/TC33 Technical Report, Nov. 1991, Draft Version 1.5.

[Fer93]   Christer Fernström. Process WEAVER: Adding Process Support to UNIX. In *[Ost93],* pages 12–26, 1993.

[FG94]   Alfonso Fuggetta and Carlo Ghezzi. State of the Art and Open Issues in Process-Centered Software Engineering Environments. *Journal of Systems and Software,* 26(1):53–60, July 1994.

[FH93]   Peter H. Feiler and Watts S. Humphrey. Software Process Development and Enactment: Concepts and Definitions. In *[Ost93],* pages 28–40, 1993.

[FKN94]   Anthony Finkelstein, Jeff Kramer, and Bashar A. Nuseibeh, editors. *Software Process Modelling and Technology.* Advanced Software Development Series, Research Studies Press/John Wiley & Sons, 1994. ISBN 0-86380-169-2, 362 p.

[Fol83]   J. D. Foley. Managing the Design of User Computer Interface. *Computer Graphics World,* pages 47–56, December 1983.

[GJ92]   Volker Gruhn and Rüdiger Jegelka. An Evaluation of FUNSOFT Nets. In *[Der92],* pages 196–214, 1992.

[GR83]   Adele Goldberg and Dave Robson. *Smalltalk-80: The Language and its Implementation.* Addison-Wesley, 1983. 714 p.

[Hø94]   Geir Magne Høydalsvik. Programmering Prosjektarbeid, Forslag til Prosjektoppgave. (In Norwegian, Working note for PhD thesis), August 1994.

[IS94]   IEEE-Software. Special Issue on Measurement-based Process Improvement. *IEEE-Software*, July 1994.

[JC93]   M. Letizia Jaccheri and Reidar Conradi. Techniques for Process Model Evolution in EPOS. *IEEE Trans. on Software Engineering*, pages 1145–1156, December 1993. (special issue on Process Model Evolution).

[L⁺89]   Anund Lie et al. Change Oriented Versioning in a Software Engineering Database. In *Walter F. Tichy (Ed.): Proc. 2nd International Workshop on Software Configuration Management, Princeton, USA, 25-27 Oct. 1989, 178 p. In ACM SIGSOFT Software Engineering Notes, 14 (7)*, pages 56–65, November 1989.

[Leb94]   David B. Leblang. The CM Challenge: Configuration Management that Works. In *In [Tic94]*, chapter 1, pages 1–37. John Wiley, 1994.

[Lon93]   Jacques Lonchamp. A Structured Conceptual and Terminological Framework for Software Process Engineering. In *[Ost93]*, pages 41–53, 1993.

[LSS94]   Odd Ivar Lindland, Guttorm Sindre, and Arne Sølvberg. Understanding Quality in Conceptual Modelling. *IEEE Software*, pages 42–49, March 1994.

[Mun93]   Bjørn P. Munch. *Versioning in a Software Engineering Database — the Change Oriented Way*. PhD thesis, DCST, NTH, Trondheim, Norway, August 1993. 265 p. (PhD thesis NTH 1993:78).

[Mye89]   Brad A. Myers. User-Interface Tools: Introduction and Survey. *IEEE Software*, 6(1):15–23, January 1989.

[NKF93]   Bashar Nuseibeh, Jeff Kramer, and Anthony Finkelstein. Expressing the Relationship between Multiple Views in Requirements Specification. In *Proc. 15th Int'l Conference on Software Engineering, Baltimore, MA*, May 1993.

[Obj92]   Object Management Group, 492 Old Connecticut Path, Framingham, MA 01701, USA. *OMG CORBA Common Object Request Broker Architecture – Specification*, January 1992.

[Ost93]   Leon Osterweil, editor. *Proc. 2nd Int'l Conference on Software Process (ICSP'2), Berlin. 170 p.* IEEE-CS Press, March 1993.

[Rob94]   Ian Robertson. An Implementation of the ISPW-6 Process Example. In *[War94]*, pages 187–206, 1994.

[SC92]   H. G. Sol and R. L. Crosslin, editors. *Dynamic Modelling of Information Systems 2*. North-Holland, 1992.

[Tic94]   Walter F. Tichy, editor. *Configuration Management*. (Trends in software). John Wiley, 1994. ISBN 0-471-94245-6.

[vV92]   Kees M. van Hee and P. A. C. Verkoulen. Data, Process and Behaviour Modelling in an Integrated Specification Framework. In *[SC92]*, pages 191–218, 1992.

[War94]   Brian Warboys, editor. *Proc. Third European Workshop on Software Process Technology (EWSPT'94), Villard-de-Lans, France. 274 p.* Springer Verlag LNCS 772, February 1994.

Experiments in Process Interface Descriptions, Visualizations and Analyses

David C. Carr[†]   Ashok Dandekar[†]   Dewayne E. Perry [‡]

[†]AT&T Bell Laboratories   [‡]AT&T Bell Laboratories
Naperville IL 60566 USA   Murray Hill NJ 07974 USA

Abstract. A wide variety of techniques and approaches are needed to understand and improve software development processes. The critical research problem is supporting the move from completely informal process descriptions to a form that includes parts that are at least machine processable. We discuss a series of engineering experiments on process visualization and analysis using simple process interface descriptions. This work grew out of two needs: the need to understand a process's internal structure and the need to understand and improve the process system's architecture. We discuss the various approaches we have taken to understand processes from these two standpoints: we report on the different forms of visualization we have used, both for processes in the small as well as processes in the large, and their resulting benefits; we delineate the various forms of analysis and report how they have played a seminal role in process improvement by providing the quantitative basis of both process understanding and process improvement efforts.

1   Introduction

For the past year and one-half, we have worked to improve our software process descriptions and process structures, as well as the software production processes themselves. We conducted a number of engineering experiments to determine the usefulness of a simple process interface descriptions language as the basis for visualization and analysis. The resulting tool fragments and tools suites have proved to be extremely useful in the understanding and improvement both of our processes and of our process system architecture.

The basis of this effort is a set of on-line manuals for the system development processes and subprocesses. These processes and subprocesses cover the entire range of the software production process from initial marketing interactions with customers through final customer support. They are described informally, but within a highly structured document format. The on-line manuals as a whole occupies about 102 megabytes of disc space and is accessed on the order of four to five hundred times a day.

Our experiments in process description, analysis and visualization began in response to two different needs. The first need was for visualization of the internal structure of a process and its data models to help in re-engineering various development processes; the second need was for an understanding of the effects on the overall process system architecture resulting from the merge of the development processes of two organizations that produce related products.

The engineering experiments described in the paper test the hypothesis that a very simple process interface description language can provide a significant amount of benefit in understanding processes both by visualization and by analysis. The results of our work have confirmed that hypothesis. Moreover, like a really useful theorem, our approach has provided the basis for a significant number of process improvement efforts.

In the subsequent sections, we describe our experimental approach, present our approach to simple process interface descriptions, discuss and illustrate our various forms of visualization, and delineate the various kinds of analysis that augment the visualization experiments. Finally, we discuss related work and summarize what we have accomplished, what we have learned from these exercises, and where we are going.

2   Experimental Approach

Experimental approaches to building software come in various guises. At one end of the spectrum is the completely rigorous experiment with attention to experimental design, well-defined experimental instruments and metrics, experimental subjects and analysis (as an example see Perry, Staudenmayer and Votta [12]). With this approach, one expects to produce quantified results about the software and its intended use. The primary purpose of scientific experimentation is to determine causality among variables in the experiment. At the other end of the spectrum are engineering experiments. With this approach, one expects informal judgments about the intended use. The primary purpose of engineering experiments is to determine usefulness. That is, we want to establish the appropriateness and utility of the experimental software and the insight it provides into the problems in question.

We take an engineering experimental approach using two complimentary methods. The first is the method on constructing the experimental software. We could call it "rapid prototyping", except that we are not prototyping but producing a complete tool sufficient for the solution needed. However, the emphasis is on rapid construction and evolution. We do this by using shell scripts to glue various existing tools together such as `dot` (a graph drawing tool [7]), sort, join and various tool fragments that operate on various transformations of the process descriptions that have been manipulated by means of pattern-matching scripts in `awk`. The various tools and tool fragments were constructed in anywhere from a few hours to a day or two. This tool-fragment approach enabled us to experiment with a variety of different approaches to both analysis and visualization without a heavy investment in system building.

The second is the method of designing, evolving and evaluating the experimental tools and tool fragments. We did this in a cooperative, incremental and iterative manner. The cooperation was between the users and the builder, proceeding iteratively in much the same manner as in a series of experiments where hypotheses are tested and evaluated and new hypotheses formed to be tested in new experiments. We proceeded incrementally adding aspects and features to the various tools as we learned what worked and what did not.

The result of this experimental approach has been a suite of related visualization and analysis tools and tool fragments that have been seminal in guiding the understanding of our existing processes and their architectural structure and that have been essential in providing the basis for process and process architecture improvement.

3   Simple Process Interface Descriptions

The basis of our experiments in visualization and analysis has been a simple form of process interface description — that is, a simple approach to describe process interfaces that is accessible to non-programmers as well as to programmers. Some of the people working on process re-engineering effort were not programmers, but were rather process domain experts. A description format that was simple and straightforward was necessary to enable them (the non-technical people) to build and understand their process and data models. Surprisingly, this approach has also been extremely effective for more sophisticated users as well.

From the standpoint of analysis and visualization experimentation, this descriptive approach has proved to be an exceedingly useful basis for quickly building sets of tool fragments. We have built visualization tools for the internal structure of process, including their associated data models, as well as for the architectural structure of the entire set of processes and subprocesses. Furthermore, we have provided analysis suites for task, subprocess, process, and cluster interfaces, various forms of cost and interval analyses, data model redundancy, and others.

Our simple description language (which we call Mini-Interact; see [10] for a description of Interact) consists of records with keywords (either explicit or implied) and one or more fields all separated by tabs. For example, a process interface description consists of a process identifier record followed first by one or more input records and second by one or more output records. The input records consist of the input artifact name and the supplier name; the output records consist of the output artifact name and the customer name. Note that this form of description remains the same whether it is a task, subprocess, process or processcluster. The only difference is the initial keyword.

TAB is used as the field separator: ⇒

Proc: ⇒ *process/subprocess-name*
Input: ⇒ *input-name* ⇒ *supplier-name*
⇒ *input-name* ⇒ *supplier-name*
. . .

Output: ⇒ *output-name* ⇒ *customer-name*
⇒ *output-name* ⇒ *customer-name*

. . .

```
Proc:    process/subprocess-name
Input:   input-name    supplier-name
         input-name    supplier-name
              . . .
Output:  output-name   customer-name
         output-name   customer-name
              . . .
```

Thus a process or subprocess interface might look like the following example of a system architecture subprocess.

```
Proc:    System Architecture Subprocess
Input:   Request                 FEP Overview Process
         Request                 Requirements Process
         Request                 Project Management Process
         Architecture Checklist  Architecture Checklist Subprocess
         Architecture Volume     System Architecture Subprocess
Output:  Architecture Volume     Requirements Process
         Architecture Volume     Software Design Process
         Architecture Volume     Hardware Development Process
         Architecture Volume     System Architecture Subprocess
         New/Updated Feature     FEP Overview Process
```

In transcribing the initial set of process interfaces, we noticed the following things about the artifacts and their suppliers and customers. First the artifacts were of widely varying granularity, ranging from a request (as in the example above) to the entire system. Obviously, this is an important architectural consideration that must be addressed in the ensuing improvement efforts.

Second, there was a wide range of characterizations of these artifacts, ranging from whole artifacts to pieces of artifacts and from plain descriptions to heavily qualified descriptions:

– Whole versus Parts — one might find the *Design Document* referenced as output from one process and the *Low Level Design Chapter* referenced as input by another.
– Generic versus Particular — a *Modification Request* is the generic form while a *Product* Modification Request is a particular kind of MR;
– By Characterization — Software Modification Requests *from previous releases* ; and
– By State — an *Open* Modification Request.

As we will see below in the analysis section, these various ways of referring to essentially the same artifact causes problems. As a result, we extended the artifact notation to describe a base artifact and to qualify it with one or more characterizations. In this way, we have a formal way of defining the base artifact and of qualifying it in a standard way. For example, the artifacts mentioned above would be described as follows.

Design Document
Design Document: Low Level Design Chapter
Modification Request
Modification Request: Product
Modification Request: Software; Previous Release
Modification Request: Open

Third, the suppliers and customers ranged over a surprisingly wide variety of possibilities. For example, we found the following as suppliers and customers:

- Processes and Subprocesses — these can be checked against the of official list of processes and subprocesses;
- Specific Roles and Classes of Roles — a specific role might be an *end user* where a class of roles might be the *product users* ;
- Organizations/Groups by Name/Descriptions — *Systems Engineering* is a specifically named organization whereas *account teams* are groups characterized by a description; and
- Collections of Roles and Groups — *field course participants* and *AT&T Business Units* are collections of roles and groups, respectively.

It is our observation that, except for a few cases such as customers, only processes and and subprocesses should be the suppliers or customers of process artifacts. We offer the following reasons for this. First, only processes and subprocesses are defined. One can look at the descriptions of the process to find out how the artifact is produced or consumed. With roles and organizations it is not possible to determine this. Second, people in the context of roles and organizations are executing some process. It is by means of these processes that roles and organizations produce and consume artifacts. Finally, the distribution of process artifacts to those executing a particular process is a matter to be decided by the that process definition itself, and not by external processes. It is a matter of encapsulation, of abstraction, of the appropriate separation of allocation and scheduling concerns.

4 Visualization — Comprehension

The initial form of visualization was needed as a part of a process reengineering effort. Our goal in this form of visualization is to aid in the comprehension of a single process — that is, to understand the relationships among the various task steps, among the tasks and their artifacts, and between the artifacts.

Initially, we differentiated the various elements in the process by different shapes and different styles of lines: ellipses represent task steps, diamonds represent external processes, boxes represent artifacts some of which may be decomposed into component parts, solid arrows indicate the input of an artifact to a task step, and dashed arrows represent the output of an artifact from a task step. The example in Figure 1 illustrates this early form of process visualization.

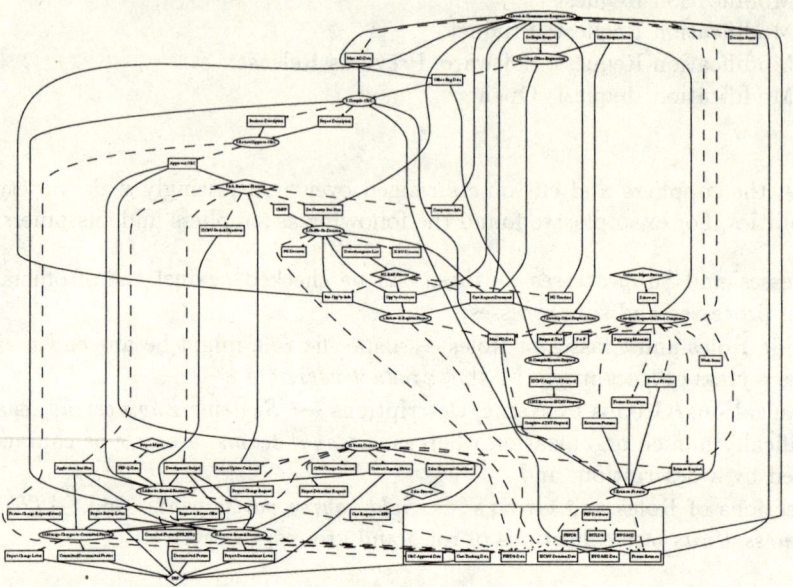

Fig. 1. Figure 1: An Example Process Visualization

The next step towards a more complete comprehension was the separation of the artifact structure from the task model. The desire of the users was to model the data separately from the tasks that manipulated it. We then extended the artifact description language to provide the typical kinds of data relationships: component parts, alternative parts, derived parts, shared components, n-ary relationships between components, etc. Furthermore, the attributes of each data component in the model were shown as subcomponents of the component. Derivation is indicated by a dashed line; sharing is indicated visually by multiple arrows, etc.

Having the capability to define various data models which interact with each other, we then provided the ability to compose data models together for both visualization and analysis purposes. Composing the models provided a visual means of determining what was common among the models. We also provided commonality analysis to provide the same information without drawing the models.

An important part of our support for comprehending processes is our desire

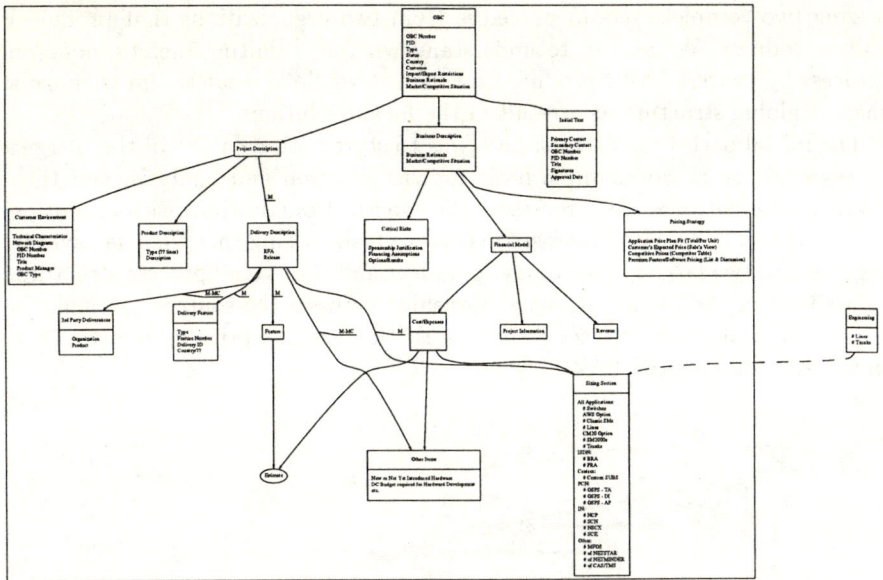

Fig. 2. Figure2: An Example Data Model Visualization

to simplify and improve those processes. An important part of the necessary background for process understanding is the cost associated with various process components, be they processes, task steps or artifacts. We added auxiliary descriptions to define various kinds of cost data that could be associated with those process components. See the analysis section below for a more complete discussion of the cost analysis. In one extension of our task visualization we extended the descriptions to incorporate the cost data into the visualization of the task model.

Further steps in the composition of the data and task models have led to the ability to color various components of the process models. We do this in two steps: first we characterize the components in a separate table. For example, we might characterize task steps as adding value to the customers or to the business, or as adding no value. We then color these different characterizations. This enables us to visually distinguish various aspects of the models.

All in all, the data and process model visualization has been very successful in several different re-engineering and improvement exercises. We continue to experiment with various ways of visually clarifying the relationships among the objects in both data and process models.

5 Visualization — Architectural Complexity

The various experiments at visualizing the internal structure of processes led to the experiments to visualize the architectural structure of the entire set of software development processes. The immediate trigger of this direction came from merging two complete sets of processes from two organizations that produced similar products. We needed to understand whether resulting merges, done on a process by process basis, produced a coherent whole — that is, did we have a coherent global structure as a result of the local evolutions.

The initial path that we took here was to start with a subset of the merged processes to use as an example basis for visualization and analysis, and then moved to the entire set of processes. We modified our initial task description syntax slightly to refer to processes instead of tasks. Using this amended syntax we generated two large graphs to try to understand the overall process structure. Figure 3 shows the input output relationships between the sources and sinks in the system. With the base set of processes, this graph's primary message was "significant complexity".

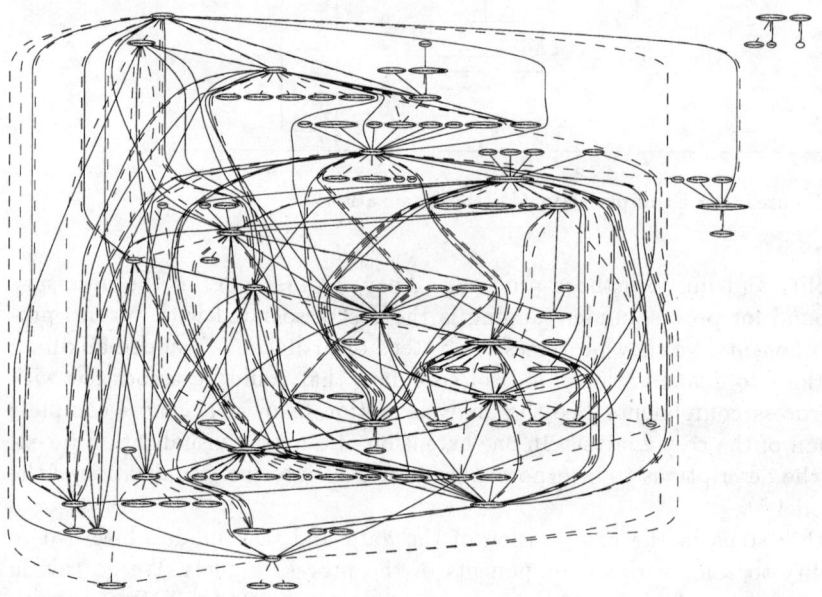

Fig. 3. Figure 3: Input/Output Relations among a Subset of Sources and Sinks

We then generated a visualization of the artifact relationships among only the processes and subprocesses. Figure 4, while still clearly indicative of complexity, was of much more use in gaining an understanding of the overall structure. This view became our canonical "big picture" of the process architecture in which the entire structure is presented at once. With the full set of processes and subprocesses, the big picture requires a graph that is 6 feet high and 25 feet long and even then the print is quite small (this is partly because long process and artifact names, but for the most part because of the sheer volume of processes and artifacts).

The interesting thing about this big picture, which clearly indicates that

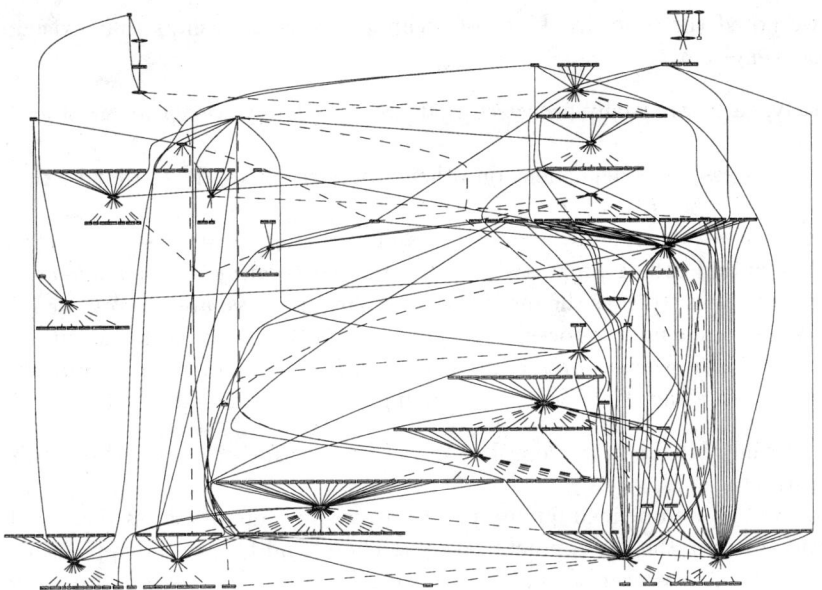

Fig. 4. Figure 4: A Subset "Big Picture" — Processes, Subprocesses and Artifacts

things are far to complex, is that it has been an amazingly effective catalyst for various important process improvement efforts. First, it is effective in showing the importance of a more coherent process architecture. Second it is effective in generating useful observations about the current architectural structure. And third, it has been fundamental in showing that we do not have an effective mechanism for managing process-in-the-large problems.

Having been encouraged by Al Barshefsky to "build it and they will come", we built the big picture and they did in fact come to look at it. The visitors ranged from developers all the way through the top levels of management. Some of the more cogent observations that came from these visits are listed below. Needless to say, these observations have provided some of the foundations for subsequent improvements.

- There is no clear path through the project.
- Project management is the center of the picture.
- A large number of artifacts are consumed internally.
- The core software development processes (requirements through coding and testing) have the simplest interfaces.
- The core software development processes appear to be sequential.
- There is a wide variance in the level of detail used to describe artifacts.
- There are an enormous number of input/output mismatches.
- There is a very high degree of fan in and fan out.
- There are no feedback loops.

In looking at the overall set of processes and how they interact with each other, we noted the following kinds of architectural relationships that exist in this process system.

- Clearly, there are many processes that are completely *independent* of each other.
- Some processes are dependent on other processes for input — that is, they are *artifact dependent*.
- Some processes cooperate with others and mutually iterate (as for example do design and coding processes) — they are *cooperating and concurrent*.
- Some processes, such as the modification request process, are *subroutine like* — that is, the current process is suspended and the MR process executed.
- Some processes, such as project management, function more as a *monitoring process* than as an dependent, cooperating or subroutine-like process.

See [4] and [11] for more complete discussions about process architectural considerations.

As with most well-defined quality programs, we have quality gates that mark important states in the development process. These tend to give the appearance of a waterfall process even thought everyone recognizes that there is far more going on both iteratively and concurrently. These gates, however, do represent some agreed upon state of either the process or the product and are important in determining progress in a project.

To try to get a better feel for what kinds of things must be going on concurrently, we provided a process dependency tree. While this is a static representation of the processes, it provides an important reminder for the large amount of work that must precede any particular quality gate in the process. Figure 5 shows the processes that must have taken place for a sample process to be able to begin: the sample process requires input from 5 processes which in turn require input, etc. We have terminated circularities in order to create a finite tree. An interesting observation derived from creating several samples was that their general appearance was not all that much different. The lesson here is that production processes, especially those considered to be the core processes, are significantly dependent on a large number of supporting processes.

6 Analysis

The visualizations gave us abundant evidence that the process interfaces were not consistent with each other. While each process interface description has both input/supplier and output/costumer sections, it was clear that these descriptions did not match in many cases.

Using the same descriptions that we used to generate the process visualizations, we incrementally evolved a set of analysis and summary tool-fragment suites to perform process interface analyses and summarizations.

- Input/Output consistency/mismatch analysis
- Process and artifact summary analysis

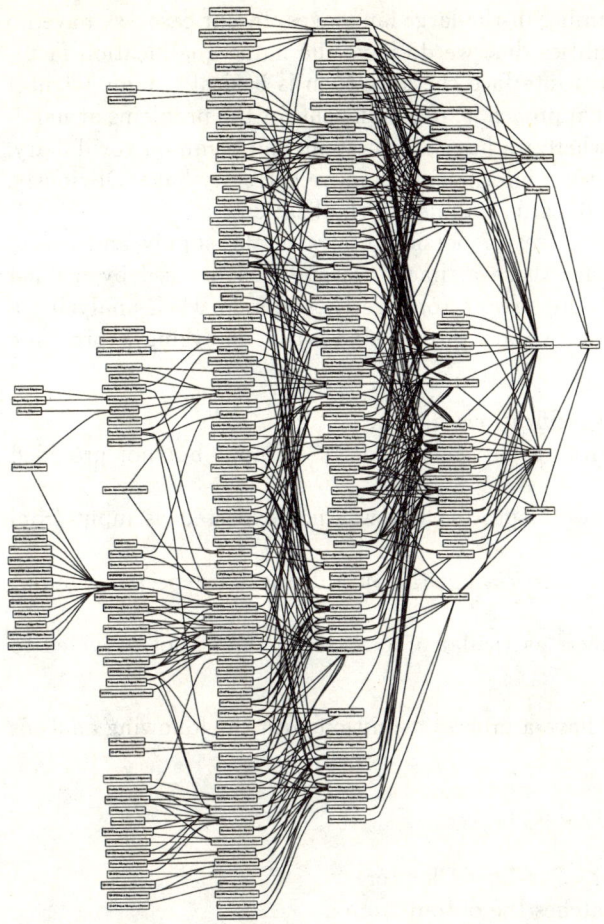

Fig. 5. Figure 5: A Sample Process Dependency Tree

- Artifact definition/usage summary
- Summary of undefined processes/subprocesses referenced
- Alphabetized artifact and source/sink lists
- Artifact and Source/Sink reference count lists and summaries

While it might be noted that we only needed the input and output artifacts to generate the pictures that we have used so far, it is in the analysis that the utility of the redundant supplier and customer information becomes obvious. Since connecting elements [13] in process architectures may be entirely informal (as opposed to the formal and automated connecting elements in product architecture), processes must be much more proactive in distributing artifacts. For this reason, the redundant information is particularly important.

The primary analysis on process interfaces checks whether the input and output specifications match. This is exactly analogous to signature checking in programming and programming-in-the-large languages. In our case, we have no formal name definition facilities, but we do have the name qualification facilities described above. Process interface checking then is basically name (string) matching. As simple as this approach is, it is sufficient for the problems at hand. Once we get to the point where we have a consistent and common vocabulary, we will then need more interface information than we currently have. Obviously, we then take the next step towards complete formalization.

We delineate the complete set of possible errors in the supply and use of process artifacts and whether those artifacts are supplied or used by defined or undefined processes. The first two errors are those found when analyzing a process's input and the last two are those found when analyzing a process's output.

- The input and output specifications match.
- Output was expected from a particular process as input but not provided by that process.
- Output was provided from a process but was not expected as input from that process.
- Output was provided to a process but the output was not expected as input by that process.
- Output was expected by a particular process as input but not provided to process.

For example, We might have a process definition with the following analysis results.

> System Architecture Subprocess
> Input:
>> Architecture Checklist Chapter
>>> matches the output from
>>> Architecture Checklist Subprocess
>> Architecture Investigation Request
>>> provided from but not expected from
>>> Requirements Process [defined]
>> Request
>>> expected from but not provided by
>>> IS FEP Overview Process [undefined]
>>> expected from but not provided by
>>> Requirements Process [defined]
> Output:
>> Architecture Volume
>>> provided to but not expected by
>>> Overall Hardware Development Process [undefined]
>>> matches the input to
>>> Software Design Process
>> New/Updated Feature
>>> expected by but not provided to

FEP Overview Process [defined]

These same inputs and outputs are summarized in the definition and usage summary in the following way.

> Architecture Checklist Chapter
>> Defined in Processes:
>>> Architecture Checklist Subprocess
>> Used in Processes:
>>> Architecture Checklist Subprocess
>>> System Architecture Subprocess
> Architecture Investigation Request
>> Defined in Processes:
>>> Requirements Process
>> Unused
> Architecture Volume
>> Defined in Processes:
>>> System Architecture Subprocess
>> Used in Processes:
>>> Requirements Process
>>> Software Design Process
>>> System Architecture Subprocess
> Request
>> Undefined
>> Used in Processes:
>>> Architecture Checklist Subprocess
>>> System Architecture Subprocess

We produce the summary of undefined processes and subprocesses referenced by comparing all customers and suppliers that have the term process or subprocess in them with the official list of processes and subprocesses. The alphabetized list of both artifacts and customers and suppliers is useful for finding those cases where mismatches are due to simple spelling errors. The reference count lists and summaries are used in various ways. The most obvious is to find those elements that are referenced only once — that is, that are either undefined or unused. Another is to find those customers and suppliers that are the most referenced. Given the observation from looking at the big picture that project management is the center of the process architecture, it is not surprising to note that project management is the most referenced customer and supplier by a factor of two over the next most referenced.

The process and artifact analysis summary provides summary data on the number of processes, inputs and outputs, the breakdown on the customers and suppliers, the number of artifacts, how many are defined and used, and the summary of the input/output analysis. We note in passing that the minimum numbers of inputs and outputs is not particularly meaningful here as they reflect some processes that were named but for which no inputs or outputs were defined. The sample below reports the summary of an arbitrary subset of the system processes.

Total number of inputs:	301
Total number of outputs:	273
Average number of inputs:	12.04
Average number of outputs:	10.92
Minimum number of inputs:	2
Maximum number of inputs:	35
Minimum number of outputs:	1
Maximum number of outputs:	35
Total distinct sources/sinks:	206
Total distinct process references as source/sink:	99
Total distinct subprocess references as source/sink:	10
Total distinct other references as source/sink:	97
Total Artifacts:	423
Total Undefined:	229
Total Defined:	194
Total Defined but Unused:	178
Total Defined and Used:	16
Total number of matches:	7
Total number of expected-from:	300
Total number of provided-from:	26
Total number of expected-by:	272
Total number of provided-to:	14

Two interesting facts are represented in this summary: first, the large number of non-process or subprocess references as suppliers and customers; second, the exceedingly large number of undefined and unused artifacts. We did a manual analysis of the sources and sinks with the following results. Of the references to processes and subprocesses, approximately 40% were correctly named, 20% were slight variants, 20% were completely misnamed, and 20% were unknown or exhibited some other problem. Of the remaining sources and sinks, about one third were possible references to processes (though they did not contain the term "process" or "subprocess"), 45% were references to roles, 35% were references to organizations.

With artifacts, since we do not have a canonical list of agreed upon artifacts as we do for processes and subprocesses, we can say only that about 10% of the non-matches are due to variants in spelling. This still leaves a large number of undefined and unused artifacts unaccounted for. We conjecture that these are the results of local definitions and a lack of global agreement on the necessary artifacts for the process system.

In summary, we have a set of analyses and summaries that can be applied to the entire set of processes, to a subset of them, or to a single process viewed in the context of the entire set. These analyses have provided us with a quantitative base of data upon which to build a successful improvement program as well as the necessary tools to aid in the improvement processes itself.

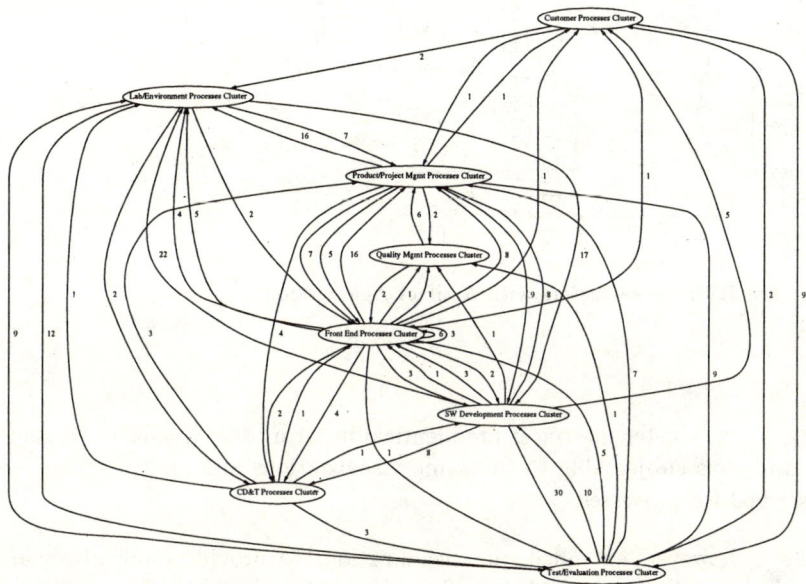

Fig. 6. Figure 6: Process Clusters with I/O Counts

7 Visualization — Architectural Abstraction

One of the main results of both the big picture visualization and the process-in-the-large analysis has been an effort to bring coherence to this process architecture. We are in the process of doing this by partitioning the processes and subprocesses into domain related clusters. The purposes of this clustering are threefold: first to partition the sheer volume of process descriptions into manageable chunks; second to create a useful architectural abstraction of the entire set of processes; and third to divide the improvement effort into that within clusters and that among clusters.

On the basis of a process to cluster mapping table and the process interface descriptions in Mini-Interact, we automatically generate the process cluster

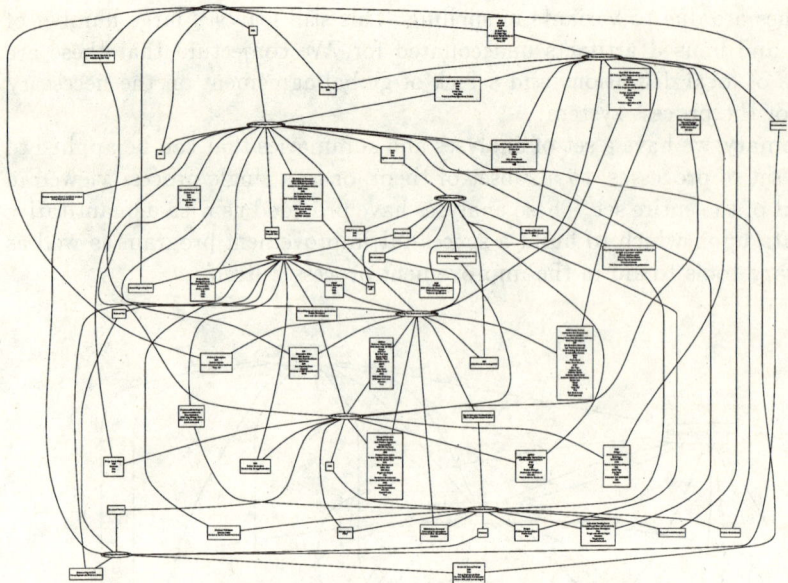

Fig. 7. Figure 7: Process Clusters with Artifact Aggregation

interfaces. These cluster interfaces are identical in form to their subsumed components and thus subjectable to the same visualizations and analyses that we have described for processes.

We have, however, extended our visualizations to provide more of an architectural abstraction approach. The first cluster visualization (Figure 6) was similar to the input/output relations that we used for the subset (Figure 3) with the extension of labeling the input and output arcs with the number of artifacts that are input or output between those two process clusters. Given that we have now on the order of 10 clusters, this makes a very understandable architectural abstraction.

The second cluster visualization expands the initial cluster visualization by providing a list of the artifacts as part of the arc between clusters. Thus we aggregate the artifacts that are passed between clusters (Figure 7) and get a more comprehensible architectural view of the clusters and their input/output artifacts.

We note that in generating these cluster visualizations we have ignored all the sources and sinks that are not defined processes in order to remove some of the clutter from the picture. This decision is defensible on the basis that, as we mentioned earlier, almost all of the input and output ought to be between processes, or in this case, between clusters.

8 Related Work

There are several different strands of related work. One strand is that of work-flow tools (of which AT&T GIS' ProcessIT [14] is a representative example). These tools generally take an interactive graphical approach and are very useful in modeling workflow as static tasks and products. They tend to have hierarchical or decomposition facilities as well. Thus at one level, there is a fair amount of similarity to the approach that we have reported here. However, there are substantial differences. First, our approach is based on an underlying textual process description from the visualizations, analyses, and clusterings are generated. This provides us an inherently more flexible basis for our experimentation. Second, we have yet to see whether these tools, as with most of the current process support tools, scale to the magnitude that we have represented here. We know of no process related tools that are set up to be used by multiple people working in a distributed manner to build a large set of interacting process or workflow descriptions.

Another strand is that of process-in-the-large facilities in various process formalisms, such as those in SLANG [3], EPOS [8], ALF [9] and APPL/A [15]. These approaches provide various ways to represent activities and use activities to build hierarchical structures. Interact [10] also has a notion of activity. However, process activities are related to each other in a non-classical way: they are related to each other by means of the policies that are used in the preconditions and the results — that is, there is an implied partial ordering of these activities. Interact further has a notion of a parameterized model in which models and activities can also appear as arguments. None of these formalisms has been used on large process systems. Thus, while there are good arguments for their utility as applied to large interacting processes, none have as yet been tested. Moreover, none give us the flexibility to address the various architectural relationships among processes that we delineated above.

The last strand is that of applying process support mechanisms to very large process systems. The only other reported account of such an effort is that of the PROMESSE project [1] using PROCESS WEAVER [6]. Their model consisted of three components: a product model, a description of process roles, and a set of lifecycle phases and subprocesses. Our work reported here supports the first and last of these aspects: both explicit and implicit product models, and of course, processes and subprocesses. There is a similarity in intent: create a specification of existing processes in order to understand the process system. The difference between the two lies in what is done with that specification. In the PROMESSE project, they implemented the specification in PROCESS WEAVER and were then able to validate it with various scenarios. In our project, we used the specifications as basis for analysis and visualization. One of our plans is to expand our cost data together with the specifications to build a queueing model with which to do various kinds of simulations (in effect, validation).

The first important difference in our approach from all of the above is the ability to generate various levels of interface descriptions, ultimately on the basis of the bottom level task descriptions. From those task interfaces, we generate process and subprocess interfaces, and from those interfaces we generate cluster interfaces. The second important difference is the interface analysis. We believe that we are the first to apply interface analysis to processes-in-the-large.

9 Summary and Future Directions

Our work has resulted in a number of different research contributions. First, we have provided an easy to understand and use process interface description language which we call Mini-Interact. With these descriptions and existing tools as the basis, we have established an effective tool creation and evolution paradigm.

Second, this paradigm has been the basis for our experiments in process visualization, analysis and interface generation. While we would not claim that our visualizations are unique (similar visualizations may be found in software understanding, for example), we do claim that we have found several that provide various kinds of insights needed to understand both interface relationships and architectural structure. Moreover, as we mentioned, we believe that we are the first to provide process-in-the-large interface analysis.

Third, we have provided insight into various process interface problems such as the various kinds of customers and suppliers referenced in the interfaces, the structure of process artifacts, and the relationships among processes.

Fourth, through our experimentation, we have demonstrated the effectiveness in our approach to supporting process understanding and comprehension. In addition to their immediate utility, they are being used as the basis for a variety of important process improvement efforts.

Three of the most important of the various improvement efforts that have arisen from this work are the following. First, we have successfully completed a process architecture study looking at the sources of our architectural problems, their underlying root causes, and various countermeasures [4]. Second, we are in the midst of several process simplification experiments in which we use various techniques to simplify and improve a complex subset of our processes [5]. Finally, we have launched an overall process system improvement effort based on the process clusters.

Our future plans include 1) incorporating the visualization and analysis as a standard part of the management of process improvement and evolution; 2) providing a graphical and interactive front end (in Xtent [2]) that will enable one to explore the architectural structure and correlate various analyses with that visual structure; and 3) include various interval, cost and quality measures to aid in deriving a queueing model of the system or components of the system for various kinds of analysis and simulations.

Acknowledgements

Mary Ellen Biell and Cindy Stach were instrumental in the early experiments with process and data modeling. Al Barshefsky has been unflagging in his support of the architectural visualization and analysis. Mary Zajac provided many of the insightful observations about the big picture. Larry Votta as a colleague in other process related experiments has provided insights for this work as well.

References

1. J-M. Aumaitre, M. Dowson, and T-R. Harjani: Lessons Learned from Formalizing and Implementing a Large Process Model. Third European Workshop on Software Process Technology, Villard de Lans, France, February 1994, Springer-Verlag.
2. Doug Blewett, Scott Anderson, Meg Kilduff, and Mike Wish: X Widget Based Software Tools for UNIX. USENIX Winter 1992, January 20-24, 1992, pages 111-123.
3. S. Bandinelli, A. Fugetta and S. Grigolli: Process Modeling In-the-Large with SLANG. Second International Conference on the Software Process, Berlin, Germany, February 1993, IEEE Computer Society Press.
4. Ashok Dandekar, Dewayne E. Perry: Experience Report: Barriers to an Effective Process Architecture - Extended Abstract. AT&T Software Symposium, Holmdel NJ, October 1994.
5. Ashok Dandekar and Dewayne E. Perry, and Lawrence G. Votta: Experiments in Process Simplification - Extended Abstract. AT&T Software Symposium, Holmdel NJ, October 1994.
6. Christer Fernstroem: PROCESS WEAVER: Adding Process Support to UNIX. Second International Conference on the Software Process, Berlin, Germany, February 1993, IEEE Computer Society Press.
7. E.R. Gansner and E. Koutsofios and S.C. North and K.P. Vo: A Technique for Drawing Directed Graphs. IEEE-TSE 19:3 (Match 1993).
8. C. Lui and R. Conradi: Process Modeling Paradigms: an evaluation. First European Workshop on Software Process Technology, Milan, Italy, May 1991, Springer-Verlag.
9. F. Oquendo, J. Zucker and P. Griffiths: The MASP Approach to Software Process Description Instantiation and Enaction. First European Workshop on Software Process Technology, Milan, Italy, May 1991, Springer-Verlag.
10. Dewayne E. Perry: Policy-Directed Coordination and Cooperation. Seventh International Software Process Workshop, Yountville CA, October 1991.
11. Dewayne E. Perry: Issues in Process Architecture. Ninth International Software Process Workshop, Airlie VA, October 1994
12. Dewayne E. Perry, Nancy A. Staudenmayer and Lawrence G. Votta: Understanding Software Development Processes, Organizations and Technology. IEEE Software, July 1994.
13. Dewayne E. Perry and Alexander L. Wolf: Foundations for the Study of Software Architecture. ACM SIGSOFT Software Engineering Notes, 17:4 (October 1992).
14. Process IT Product Brochure. AT&T GIS.
15. S. Sutton, D. Heimbigner and L. Osterweil: Language Constructs for Managing Change in Process-Centered Environments. Fourth ACM SIGSOFT Symposium on Software Development Environments, Irvine CA, December 1990. ACM SIGSOFT Software Engineering Notes, December 1990.

Related Domains (Chair: Colin Tully)
The Software Process and the Modelling of Complex Systems

Colin Tully

Colin Tully Associates, Tunbridge Wells, UK
and
Division of Computer Science, University of Hertfordshire, Hatfield, UK

1 Systems, Models and Complexity

Scientists and engineers (and others) build *models* primarily to gain understanding, of existing or future phenonema, in natural, artificial or social *systems.* The understanding to be gained from models concerns essentially how the *behaviour* of systems, subject to a range of external stimuli, is governed by their *structure.* That understanding may simply serve to satisfy scientific curiosity, or it may go further and serve as the basis for subsequent planning and action. (It may be necessary to undertake further modelling in order to try and understand, or predict, the effects of the actions that are planned.)

Of course, there are all kinds of models that might be made. As scientists and engineers, we are interested in *effective models,* where the criterion of effectiveness is the degree of understanding of a system, related to the purposes in hand, to which a model leads.

What is meant by treating any subject as a system S is that it comprises not only a set P of *parts* but also a set R of *relationships* among the parts [Kaposi 1994]. The inclusion of the set R of relationships is critical, and it underlies the common wisdom that, for instance, a system is more than the sum of its parts, or that the concept of a system is different from (and richer than) the concept of a set — though, as we have just seen, the system concept can be expressed in terms of the set concept. When we talk of the structure of a system, we mean precisely the pair (P, R).

Systems vary in the *complexity* of their structures. Indeed the complexity of systems is something we should dearly like to be able to measure, in absolute or relative terms, though there are no convincing general proposals. (Murray Gell–Mann, the Nobel physicist, is among those who have recently addresssed the issue [Gell–Mann 1994]. While his effort is illuminating, it seems to raise at least as many difficulties as it removes.) What seems not in dispute, however, is that complexity is strongly related to the set R, and more particularly to the cardinality and variety of that set. (By variety, in this context, is meant the number of separate subsets into which the members of the set may be classified.)

If a system is to be *viable* — that is, if it is to be able to generate a range of behaviour which is adequate to handle the range of environmental stimuli — its structure must have sufficient complexity to allow that range of behaviour. This principle was originally, and memorably, expressed by Ross Ashby as the principle of requisite variety [Ashby 1956]. Of course, not any old complexity will do: it has to be "the right" complexity, permitting "the right" behaviour. Complexity for its own sake is a weakness in any system.

We know there are factors that add to system complexity, and which lead to behaviour which may be chaotic [Waldrop 1992] and counter–intuitive [Forrester 1969]. Important among them is the existence of *causal loops*.

In any system, cause–effect relationships form a class of relationships in the set **R**. Causal loops occur when a subset of parts in the set **P**, and a subset of cause–effect relationships among them, form a loop. It is often the case that causal loops give rise to exponential growth or decay (virtuous or vicious circles, or positive feedback), or to states of balance or stability (control, or negative feedback).

There are (at least) three conditions in which causal loops lead to special difficulties in making effective models of system behaviour:

- if there are long time–lags in circuiting causal loops;
- if causal loops, each with its own "time signature", interact;
- if causation is "intentional" (arising from the prior formation of human intentions) rather than "natural" (arising from the operation of the "laws of nature") [Searle 1984].

Worst is if all three conditions apply together!

It is platitudinous to observe that each of the parts of a system is itself a system, and therefore that all the above observations apply equally to the parts as to the whole. This familiar "Russian dolls" concept of systems within systems is often called the *systems hierarchy*.

Aficionados of the systems approach often rebuke scientists for *reductionism*. It is often mistakenly thought (even by the aficionados themselves) that they mean, by that rebuke, that scientists do not employ systems models. If that were true, however, scientists would not have effective models to explain the phenomena they study. What the reductionist charge really means is that science tends to stop enquiry at too low a level in the systems hierarchy, and to concentrate excessively on details at the lower levels. There is, of course, a plausible reason for that: as you get higher in the hierarchy, systems tend to become more complex, because the relationships become more numerous, various and interrelated, and in particular because causal relationships display the conditions listed above. In these circumstances, it becomes harder and harder successfully to employ established scientific methods, and scientists prefer to confine their effort to the lower–level phenomena which their established methods can handle.

2   Modelling the Software Process as a Complex System

I propose to begin this second section with an assertion that I hope can stand undefended. It is that real software processes, in real organisations, are complex systems. We cannot put numbers on that complexity; but if you ask "how complex?", the answer is "very".

In passing, I should like to propose an illustrative hypothesis. It is that any individual real software process is more complex than its product — and we are accustomed to priding ourselves that software products are, as a class, the most complex of human artifacts. In the absence of accepted measures of system complexity, that hypothesis cannot be rigorously tested. Informally, however, the following argument may be offered. Consider a single work product: say a requirements specification. Intuition suggests that the (sub)process by which the specification is produced is more complex than the specification itself. In that case, it is reasonable to suggest that the totality of a real process as a system is more complex than the totality of its work products as a system.

Be that as it may, there is increasing urgency to understand real software processes in real organisations, so that those processes can be improved, so that in turn the host organisations can improve the quality–cost–timeliness attributes of the strategically critical software which is output by the processes. That means making effective models of software processes, both existing and planned, so as to understand how their behaviour is governed by their structure. It means in particular understanding how processes are composed of parts (subprocesses) and, critically, understanding the relationships that exist between those parts. Finally, given the high degree of complexity of software processes, it is likely that causal loops — including positive and negative feedback loops [Lehman 1994] [Weinberg 1992] — must be properly identified in models of existing processes, and must be properly positioned in models of planned processes.

Proposing methods of modelling software processes, and testing the validity and effectiveness of those methods on existing processes, is a scientific activity. Using those methods to design new processes, which will produce "better" software, is an engineering activity.

The position which it is the purpose of this paper to express is that, in both scientific and engineering terms, the software process modelling enterprise has so far been largely inadequate, either through failing to take account of complexity or through being reductionistic. Our modelling methods, and therefore the models which we try to use, are not effective for our purposes of understanding and of subsequent planning and action. That failure is serious, given the urgency of the practical need. (A similar criticism might be equally applicable to business process modelling; but that is a different issue.)

There is only space for two illustrations of that position. In the context of the first allegation, that models ignore the complexity of reality, let us consider the class of life–cycle models and, as a specific example, the waterfall model in its many variants. The parts of the process, according to waterfall models, are phases, and the relationships between them are precedence and backtracking. This is a manifestly inadequate set of modelling concepts, and experience suggests that models based on them are ineffective for understanding real processes and their strengths and weaknesses, or for designing process improvements. Their main virtue seems to lie in providing a very simple form of management appreciation, centred around the notion of project milestones.

In the context of the second allegation, that models are reductionistic, let us consider the class of models proposed by most research teams. These normally adopt modelling concepts which are already in use for modelling software products: these formalisms are too numerous to list here, and in any case they are well known. In these methods, the parts of the process are things like primitive actions, information structures, rules etc, and the relationships between them are things like precedence, concurrence, input, output, composed–of, consistent–with etc. These system concepts are fully adequate for describing mechanised processes. That is what they were developed for, and they can be successfully applied also where the purpose is to guide or control routine human activity through "process engines" (whether embedded in software engineering environments or free–standing). But that application is only relevant where the human activities are of a low–level, routine or machine–like nature — that is for low–level fragments of the whole process. Such models could only purport to explain the performance of the overall software process in the context of a real organisation if the quality of the process were thought to be determined by the extent to which human actors slavishly follow defined procedures.

We have reached the point of suggesting that, if we wish to understand the behaviour of the software process as a complex system and as part of a larger business process, and if we wish to be able to model improvements to it so that it will sustain a viable existence in response to the stimuli received from that larger system, then the modelling capabilities we can currently deploy are inadequate. There is, however, a third class of process models which we have not yet mentioned, and which may provide the basis for some hope. They are process assessment models.

Process assessment models originated from Watts Humphrey, whose original model has been developed into the current version of the SEI CMM [Paulk 1993]. Others have been proposed, such as SDCE, Trillium, SQPA, Bootstrap, Software

Technology Diagnostic etc. The most advanced model, still being developed, is SPICE [SPICE 1994]. The purpose of all those models is to decompose the whole process into a hierarchy of subprocesses, with the objective that the "maturity" of an individual organisation in performing the subprocesses (either separately or in aggregate) can be measured on a maturity scale. The purpose, in other words, is measurement not understanding. And certainly these models, as they stand, cannot be regarded as system models: they identify only the set **P** of parts (subprocesses) and not the set **R** of relationships between them.

Assessment models nevertheless have merits and potential. They avoid the reductionism of most research laboratory models by modelling the process as a whole. They avoid some of the over–simplicity of life–cycle models by identifying an extensive repertoire of parts (set **P**) of that whole process. SPICE, for instance, proposes a hierarchy of five process areas, 36 processes and 213 base practices; that ranks as a scientific hypothesis of some magnitude, which will be subjected at least to partial test during the SPICE trial period. They have the further merit of being known and accepted in the practitioner community.

How could they be the basis for the kind of explanatory models which have been argued for in this paper? The answer is to add a set of relationships **R** to the existing set of parts **P**. In doing so, particular attention should be paid to identifying types of causal loops, and to partitioning those types into ones which are beneficial and ones which are harmful [Lehman 1994].

3 Conclusion

This paper outlines the rationale for a new scientific and engineering approach to software process modelling, based on a clear understanding of systems principles. The purpose of the approach is to align the work of the research community with the practical process improvement needs of management and practitioners in industry, business and government. It is suggested that assessment models provide a basis for such an approach, even though their purpose is different and they therefore need to be developed to suit the purposes discussed in this paper.

How they could be developed is beyond the scope of this position paper (which is already too long). One thing, however, can and should be said. The trap of having mechanistic models should be avoided at all costs. One criticism of process models is indeed that they risk taking us into an era of neo–Taylorism. The organisational paradigm into which we should seek to fit them is rather that of the learning organisation, which seeks to maximise individual learning and to tap it as an asset of the organisation as a whole. It is significant that one of the seminal books on the learning organisation [Senge 1990] identifies systems thinking, including the identification and management of causal loops, as the fundamental discipline on which the learning organisation, the organisation of the future, must be founded.

References

[Ashby 1956]   Ashby W R. *An introduction to cybernetics.* Methuen, 1956.

[Forrester 1969]   Forrester J W. *Industrial dynamics.* MIT Press, 1969.

[Gell–Mann 1994]   Gell–Mann M. *The quark and the jaguar.* Little, Brown, 1994.

[Kaposi 1994]   Kaposi A and Myers M. *Systems, models and measures.* Springer, 1994.

[Lehman 1994]   Lehman M M. *Introduction to FEAST.* In Proceedings of FEAST workshop.   Department of Computing, Imperial College, London, 1994.

[Paulk 1993]   Paulk M C, Weber C V, Garcia S M, Chrissis M B and Bush M. *Key practices of the capability maturity model, version 1.1.* Technical report CMU/SEI–93–TR–25. Software Engineering Institute, 1993.

[Searle 1984]   Searle J R. *Minds, brains and science.* British Broadcasting Corporation, 1984.

[Senge 1990]   Senge P M. *The fifth discipline: the art and practice of the learning organisation.* Doubleday, 1990.

[SPICE 1994]   SPICE Project. *SPICE baseline practices guide.* 1994.

[Waldrop 1992]   Waldrop M M. *Complexity: the emerging science at the edge of order and chaos.* Simon & Schuster, 1992.

[Weinberg 1992]   Weinberg G M. *Quality software management. Volume 1: Systems thinking.* Dorset House Publishing, 1992.

Interpretable Process Models for Software Development and Workflow

Gerhard Chroust

Systems Engineering and Automation
Kepler University Linz
A-4040 Linz-Auhof
chroust@sea.uni-linz.ac.at

Abstract. Guiding software development via an enacted process model has by now become state-of-the-art, leading to Software Engineering Environments. Similarly administrative office work also largely follows pre-defined procedures, laws, and regulations, which essentially also establish a process model. Computer support for this field is currently hotly discussed under the catch word of 'work flow'.

We explore similarities and differences of both fields with respect to a variety of characteristics. It concludes that both fields obey the same basic paradigm, i.e. describing the desired processes by a process model and enacting this model by a process mechanism. The characteristics are sufficiently similar to justify a common approach, but at the same time there exist significant differences which make it necessary to use different implementations for software development and administrative processes.

1.0 Background

Mechanise the Mechanisers!

1.1 Software Engineering

The history of *software development* shows a trend to achieve an industrial maturity (in contrast to an artistic approach) by following a well-defined software process [7] [8] [14] [24]. The desired and anticipated development process is defined by a process model, e.g. IBM's ADPS [3], the German V-Model [2] or the MAESTROII-Model [18]. Enactment of the process model takes place on the basis of a software engineering environment [17] [21]. This trend has been considerably enforced by the current push for process maturity as expressed by the Capability Maturity Model [19], the Bootstrap-project [9], the ISO 9000 standard [6] and similiar approaches. The need to support individual activities by computer based tools and the necessity of a repository for results (see Figure 1) leads to *software engineering environments* [13] [18] [21].

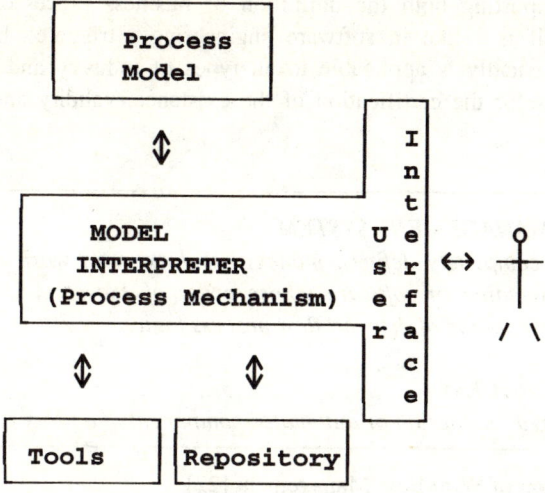

Figure 1. Components of a SW Engineering Environment

1.2 Workflow Management

Independently, and actually also much earlier, administrative procedures have existed. Formalization is inherent in many bureaucratic transactions in government ('red tape') and administration, in the banking and insurance business (typically taking out a loan, signing up for a car insurance, installing a new complex computer system etc.). Back to pre-historic times administration, then often intimately interwoven with religious practices, has been relying on established procedures of handling the daily routine and the necessary religious acts. It is generally accepted [10] that a considerable potential of productivity improvement rests in automating the office world [1]. The application of the process model/process enactment paradigm is a natural idea since the definition and observance of office procedures and business processes itself has thus a long lasting history.

Observations confirm that up to 90% of the elapsed time in some business processes is actually wait and transfer time [10]. This chance has currently brought Workflow Management (for a definition see Figure 2) into the limelight of commercial interest [1]. This is exemplified by numerous products on the market [15].

A careful study of the definition in Figure 2 allows the conclusion that work flow management is obviously only applicable to the routine type of processes occurring in an office. [11] speaks of the *'computer supported performance of strongly structured, chained office procedures of repetitive type'* , i.e. the routine-oriented work [20].

The need for supporting both the definition of business processes and the their enactment in itself is - like in software engineering - triggered by ISO 9000, a standard which basically is applicable to all types of industry and commerce and which is the basis for the certification of the existence, validity and observance of business processes.

WORKFLOW MANAGEMENT SYSTEM
A system that completely defines, manages and executes workflow processes through the execution of software whose order of execution is driven by a computer representation of the workflow process logic.

WORKFLOW PROCESS:
The computerized facilitation or automated component of a process.

Figure 2. Definition of Workflow Management [22]

2.0 Software and Work Flow

A major question is whether software development and administration are sufficiently similar such that conceptually the *same* process description methods and the *same* process enactment concepts can be applied, or whether completely different concepts have to be employed [5]. In the sequel their key properties and major differences which might affect the process mechanism will be discussed.

The comparison will only consider *typical* behavior with respect to software engineering and administration. The following criteria will be discussed. Further arguments can be found in [5]:

- Character of processes
- User's role perception
- Documentation of process
- Security considerations
- Multiplicity
- Process Enactment

3.0 Detailed Comparisons

Table 1. Character of Processes		
	software development	*workflow management*
Character of the Process	A Process Model describes a software development process, which creates an application in a systematic, high-quality way.	A process model describes an administrative procedure in order to arrive at some verdict, conclusion etc. in a systematic, high-quality way.
Applica-bility	Applicable to most software processes. Developments with a high ratio of prototyping make difficulties.	Applicable if a well-established procedure ('routine work') is to be followed. Processes with considerable freedom of action make difficulties.
Type of steps	Numerous creative steps are involved, interspersed with mechanical transformations (e.g. compilation). It is important to apply the correct transformations and to use the appropriate tools. In principle one follows the process model, deviations are usually admissible.	The steps and their sequence is fully defined in advance. For various reasons (laws, rules) they have to be performed in exactly that sequence. Deviation are discouraged, closely controlled and logged.
Semantics of results	The developed application itself is a *scheme* for processes to be performed (i.e. the execution of the application).	Process enactment is the end purpose. Results document the actions performed or the decisions reached (e.g. verdicts).

Table 2. User's Role Perception		
	software development	*workflow management*
Enact-ment	Users *develop* a description of a (second) process, i.e. the actual application. The 'creativity' of this work is often over-estimated.	Users *enact* the process. They see themselves as officials following a procedure, making the necessary decisions, based upon facts and judments.
Respon-sibility	The project manager coordinates and delegates work, subdelegation is not common.	The responsible person performs the key steps (decisions, ...) himself, subordinate activities are delegated and may be subdelegated again.
Standards	Standards for results are useful, sometimes required.	Standardized results are the rule.

Table 3. Documentation of Process		
	software development	*workflow management*
Notation	data flow type	control flow type
Need	For project management (e.g. causal analysis, costing) a log of the activities may be kept. The exact sequence of activities is of little interest.	Legal or company regulations usually require that the sequence of steps and all decisions be preserved, including the reasons leading to them.
Archiving	At the end of a project most of the project documentation has only historical interest. Only created (final) results and the applied transformations persist.	All steps and their order of performance must be retraceable. Intermediate results and versions thereof must be preserved.

Table 4. Security Considerations		
	software development	*workflow management*
authori-zation	Authorization is not very critical *within* a development team.	Authorization to perform certain activities must be strictly controlled.
Read and write	Browsing of results is usually uncritical. Creation or change of a result must be strictly controlled.	Access to all data has to be strictly controlled - even read-only access.
System failure	The *creative* contents of the created results must be preserved, mechanical steps can be re-done.	The *status* of the process must be preserved, re-start must be from exactly that status.

Table 5. Multiplicity		
	software development	*workflow management*
Trans-action rate	Many parallel activities, many developer work in parallel.	Usually little parallelism within a process, one or a small team work at one process at a time.
Number of processes	A few dozen processes at most are active at the same time.	Many hundred and more processes run in parallel.
Number of instances	For a majority of result types many (possibly hundreds) of instances exist (e.g. hundreds of code modules).	Usually the complete process model is instantiated once for each actual transaction.
Fan-out, fan-in	Considerable 'fan-out'/'fan-in' for results (e.g. a 'function' expands in many 'modules', many 'modules' belong to one 'subsystem').	Transactions are fairly 'lean' having little fan-out/fan-in. Essentially a small set of objects ('documents', 'folders') is being passed through the transaction.

Table 6. Process Enactment

	software development	*workflow management*
Sequencing	The sequence of activities is only approximately defined, data dependency being a major constraint. Sequencing is analog to data flow machines. The user usually chooses one of the 'ready' activities.	The sequence of activities is usually strictly predetermined, in the spirit of control flow similar to conventional programming languages. The user has little freedom.
successor function	After completion of an activity usually many different activities are available for continuation. Most of them belong to the same process.	For a given activity usually a small set of successor activities exists. Not much freedom of choice is available within a process, but many different processes exist.
iteration, repetition	Repeating an activity is often a ncecessity (e.g. Spiral Model, prototyping) and usually is not critical. The system must keep track of different versions of results	Each step's enactment is strictly controlled, iteration is often formally defined by special 'correction activities'. Explicit loops and refinement steps are rare.

4.0 Outlook

The comparison indicates that software development and workflow management are sufficiently similar to be enacted basically by the *same* process mechanism. When actually constructing such process mechanisms, however, one has to recognize differences, which imply different emphasis and different 'tuning'. Key differences seem to be:

Character of Process: In software the creativity is expanded in choosing the transformations and thus partially the sequence. In administrative processes the sequence is fixed, creativity is spent *within* the defined activities.

Multiplicity: In software development the number of instances of the individual types of components 'explodes in the middle' (i.e. in the phase 'implementation' there exist many 'code modules') while it is small at the 'beginning' and the 'end' (usually one 'requirements document' and one 'integrated system'), cf. [12]. Administrative processes, on the contrary, show only a very small number of instances per component of a process, the number of instances of the *whole process*, however, is often horrendous: while a few dozen software projects are probably the maximum in an organization, thousands of parallel and independent transactions, e.g. cashing checks, are the normal case.

Freedom of Navigation: In administrative processes a strict observance of the prescribed sequence is usually required and often even postulated by law. Software development itself does not require a pre-set sequence of activity. The sequence of activities is primarily constrained by data dependencies. Any additional constraints are imposed by management to foster controllability and transparency of the process but are not inherent to software development itself.

Documentation of the Process: Administrative procedures usually require a complete record of the steps taken and their outcome, including intermediate versions. [23] states that administration is characterized by *'the uninterrupted documentation of all processing steps. The resulting transparency and auditability'* is a cornerstone. In software development we are satisfied to have *strict control of a liberal process* in the sense that the *logical sequence* of the steps has to be observed and their results preserved [14]. Repetition of steps, as long as logical relationships of the results are uneffected, is permissible.

Thus it is plausible that despite certain differences in amount and rigor essentially the same process mechanism can be utilized for both software development and administration. This indicates that essentially the same descriptive means can be used to describe software processes and administrative procedures [16]. Software engineering itself partly proves above statement: The evolution of process models [4] shows a continuous expansion of the breath of such models, in the sense that continuously more accompanying actions (e.g. documentation, quality assurance, product management) are also included in the model and expressed and enacted by the same interpretative means. These accompanying actions usually behave like administrative procedures.

This outlook promises the integration of software development with other administrative processes. Hopefully this will lead to a uniform systems view of the various domains of an enterprise. The importance of this integration will prove itself in the process of certification with respect to ISO 9000 [6], where complete business processes will be analyzed. For software development departments this means certification of both some software development and some administrative processing. A common paradigm, a common language and a common process mechanism will be of advantage.

5.0 References

1. Bergsmann J.: Workflow im gewerblichen Bereich.- Diplomarbeit, Kepler Universität Linz, Sommer 1994.

2. Bröhl A.P., Dröschel W. (eds.): Das V-Modell - Der Standard für die Softwareentwicklung mit Praxisleitfaden.- Oldenbourg 1993

3. Chroust G.: Application Development Project Support (ADPS) - An Environment for Industrial Application Development.- ACM Software Engineering Notes, vol. 14 (1989) no. 5, pp. 83-104

4. Chroust G.: Modelle der Software-Entwicklung - Aufbau und Interpretation von Vorgehensmodellen.- Oldenbourg Verlag, 1992

5. Chroust G., Leymann F.: Interpretable Process Models for Software Development and Administration.- Trappl R. (ed.): Cybernetics and Systems Research 92, Vienna, April 1992, World Scientific Singapore 1992, pp. 271-278

6. Deutsches Institut für Normung: DIN ISO 9000: Leitfaden zur Auswahl und Anwendung der Normen zu Qualitätsmanagement, Elementen eines Qualitäts- Sicherungssystems und zu Qualitätssicherungs-Nachweisstufen.- Beuth Verlag 1987

7. Dowson M., Wileden J.C.: A Brief Report on the International Workshop on the Software Process and Software Environments.- SIGSOFT Software Engineering Notes vol. 10 (1985) No. 3, pp. 19-23

8. Eliadis D.: SPM, the lost component.- Software Magazine, 10. Okt. 1991, pp.11-12

9. Haase V., Messnarz R., Koch G., Kugler H.J., Decrinis P.: Bootstrap: Fine-Tuning Process Assessment.- IEEE Software vol. 11 (1994), no. 4, pp. 25-35.

10. Hammer M., Champy J.: Business Reeengineering - Die Radikalkur für das Unternehmen.- Campus Frankfurt/M, 3. Auflage, 1994

11. Hansen H.R.: Arbeitsbuch Wirtschaftsinformatik I.- Fischer Verlag 1993

12. Hruschka P.: Information Hiding Module kommuniziert gezielt.- Computer Woche, 8. April 1983, pp. 14-16.

13. Huenke H. (ed.): Software Engineering Environments.- Proceedings, Lahnstein, BRD, 1980, North Holland 1981

14. Humphrey W.S.: Managing the Software Process.- Addison-Wesley Reading Mass. 1989

15. Inst. f. Internat. Research.- Dokument Management Systeme - Eine Produkt- und Marktübersicht.- IMACO GmbH, Inst. f. Internat. Research, Zürich, Sept 1994

16. Knöll H.D., Suk W.: Eine graphische Sprache für kommerzielle Programmiersysteme.- Schriftenreihe Inst. f. Angew. Wirtschaftsinformatik, No.2, 1989

17. McDermid J. (ed.): Integrated project support environments.- P. Peregrinus Ltd. London 1985

18. Merbeth G.: MAESTRO-IPSE - die Integrierte Software- Produktions-Umgebung von Softlab.- Balzert H. (ed.): CASE - Systeme und Werkzeuge.- B-I Wissenschaftsverlag 1989, pp.213-234

19. Paulk M.C. et. al.: Capability Maturity Model, Version 1.1.- IEEE Software vol. 10 (1993) July 1993, pp. 18-27.

20. Schmidt S.: Büro-Informationssysteme - Ein Überblick.- Informatik-Spektrum vol. 12 (1989) No.1, pp. 19-30

21. Sommerville I. (ed.): Software Engineering Environments.- P. Peregrinus Ltd. London 1986

22. Glossary - A Workflow Managment Coalition Specification.- Workflow Management Coalition, Belgium Nov. 1994
23. Wiesboeck H.: Anforderungen an ein Kanzlei-Informationssystem der öffentlichen Verwaltung.- ADV (ed.): EDV in den 90er Jahren: Jahrzehnt der Anwender - Jahrzehnt der Integration.- ADV 1990, pp. 719-728.
24. Wileden J.C., Dowson M. (eds.): Internat. Workshop on the Software Process and Software Environments.- Software Eng. Notes vol. 11 (1986) No. 4, pp. 1-74

Acknowledgement. The author would like to thank Dr. Frank Leymann and Ms. Ulrike Knapp, IBM Germany, for early contribution to this paper.

Integrating Process Technology and CSCW

Elisabetta Di Nitto and Alfonso Fuggetta

Politecnico di Milano
Dip.to di Elettronica ed Informazione
P.zza Leonardo da Vinci, 32
20133 Milano - Italy
email: {dinitto, fuggetta}@elet.polimi.it

Abstract. Software processes are complex activities where designers, managers, programmers, and users must cooperate to achieve effective results. The work done in the software process community during the last years has been mainly focused on the provision of means to model software processes and to support asynchronous, text-based communication. On the other side, CSCW (Computer Supported Cooperative Work) technology aims at supporting general cooperation activities among human agents exchanging multimedia information. Recently, there has been an increasing interest on the issues related to the integration of these two technologies.

This paper discusses different integration strategies between the SPADE PSEE (Process-centered Software Engineering Environment) and the ImagineDesk toolkit for the development of CSCW systems. It outlines how the advantages and limitations of the two systems can be matched to achieve more effective computer support in software production.

Keywords and phrases: Software processes, CSCW, PSEE, cooperation and coordination.

1. Introduction

In recent years, there has been an increasing attention on supporting human activities characterized by a high degree of cooperation. This interest is demonstrated by many initiatives that have been started in different cultural contexts.

In the business administration field, there has been a huge amount of research and experimentation, aiming at improving the quality of company processes. Typical examples of such efforts are the TQM and Business Reengineering approaches. All of these approaches suggest that one of the key factors to improve the performances and results of an organization is the efficiency and effectiveness of its communication channels.

In information technology the need for communication, coming from different application field, results in a growing interest versus systems supporting cooperation. In particular, there has been an increasing attention to the development of languages and tools to explicitly describe and analyze complex processes, in which humans exchange a large amount of information. In addition, tools providing automated support to process activities and agents are being realized. Typical examples are Process-centered Software Engineering Environments (PSEEs) and workflow systems.

Moreover, the general problem of cooperation in different classes of organizations has been widely studied, and has generated the research area known as Computer Supported Cooperative Work (CSCW).

Recently, several researchers have pointed out that PSEEs and CSCW can offer complementary contributions to the general issue of providing process support to software development organizations (see for example [1]). In this paper we explore the issue, by illustrating possible integration strategies for a process-centered environment (SPADE) and an environment for building and executing cooperative applications (ImagineDesk). We believe that the results are general enough to be reused also in other contexts. In fact, both SPADE and ImagineDesk can be considered state of the art results of software processes and CSCW fields. Sections 2 and 3 provide the reader with a very short description of the architecture of SPADE and ImagineDesk. Section 4 introduces the integration strategies we have identified, and provides some observations and critical reflections. Finally, Section 5 offers some final considerations.

2. A Quick Overview of the SPADE Architecture

SPADE is a process-centered software engineering environment (PSEE) based on a language for process modeling called SLANG [3]. Its development started in 1991 at Politecnico di Milano and CEFRIEL. The architecture of SPADE is based on three separate layers (see Figure 1):

- Process Enactment Environment (PEE).
- User interaction Environment (UIE).
- The SPADE Communication Interface (SCI).

The Process Enactment Environment includes facilities to execute a SLANG specification, and to create and modify process artifacts. The User Interaction Environment manages the interaction between users and the process. The SPADE Communication Interface supports communication between the PEE and the UIE.

The PEE task is to execute a SLANG process model. The main component of the PEE is the Process Engine (PE). It is responsible for the execution of multiple SLANG Interpreter instances. Each SLANG Interpreter executes a different SLANG fragment. The PE is implemented as a Unix process, and each SLANG Interpreter is a parallel thread of execution. The PE is connected to the SPADE Monitor (SM), and notifies it of relevant events (transition firing, new SLANG Interpreters instantiations/terminations etc.). SM presents the state of the process enactment to SPADE users, by providing a comprehensive visualization of the state of execution of the SLANG fragments instantiated so far.

The goal of the User Interaction Environment is to manage the interaction between SPADE and its users. Users coordination and interaction is achieved through tools that are integrated into SPADE. Tools in the UIE can be *black-box tools* or *service-based tools* depending on the level of control integration. Black-box tools are traditional Unix processes that are launched by SPADE. Their termination is triggered by SPADE, but no other interaction with the PEE is possible during their execution. Service-based tools provide services that can be invoked through a programmatic interface. Service-based tools can be directly connected to the SCI (see next paragraph) or can belong to tool integration facilities such as DEC FUSE or Sun Tooltalk. In this latter case, tools are not directly connected to the SCI, but they communicate with SPADE through a *bridge* that translates the protocol of the specific tool integration facility into the SCI protocol.

The SPADE Communication Interface is a filter which allows communication between the PEE and the UIE. The communication is based on a message passing paradigm, and follows a precise protocol. It is a Unix process, connected with the PE and each service-based tool or tool integration facility to be used. The SCI allows the PE to invoke the services exported from the tools in the UIE, and to receive notifications of

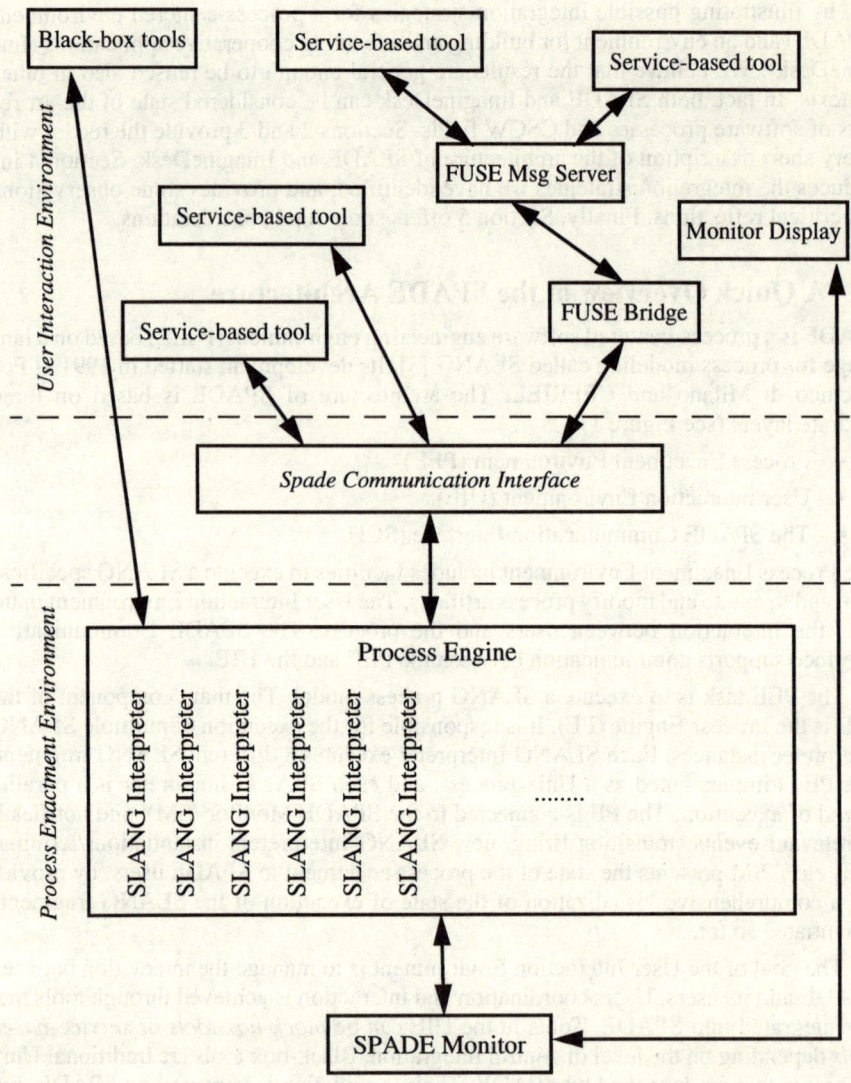

Fig. 1. SPADE Architecture.

relevant events from the UIE. These events are reified as data of some SLANG Interpreter, and can subsequently be manipulated as any other information in the process model[1].

3. A Quick Overview of the ImagineDesk Architecture

The ImagineDesk platform supplies a toolkit to build synchronous cooperative applications, and an environment to execute them [4].

---

[1] For more information on the architecture of SPADE the reader can refer to [5].

Synchronous cooperative applications simulate a workspace that is shared by users to communicate. They have to offer some functionalities to manipulate the content of the shared workspace and to control the access to it, in order to avoid that users perform conflicting actions.

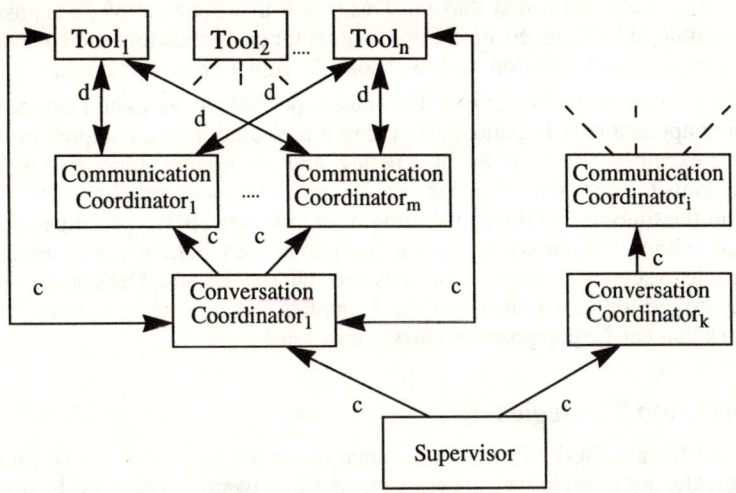

Fig. 2. The ImagineDesk applications architecture.

An ImagineDesk cooperative application presents the distributed architecture shown in Figure 2. In the figure boxes represent processes, while arrows are used to specify communication channels. A "c" or "d" near to an arrow indicates that the channel is used for control or data transmission. Each process composing the architecture is devoted to perform a specific task:

- *Tools* implement the user interface of the cooperative application. For each user that is attending the cooperation, a tool is executed. It allows the user to have a view of the shared workspace, to perform actions on it, and to change his/her role in the cooperation.

- *Communication Coordinators* are devoted to control the flow of the data among tools. If the application manages different types of media (e.g. images and text), different Communication Coordinators, one for each type of media, are used. They are connected to all the tools through bi-directional data communication channels, and are controlled by a Conversation Coordinator.

- The *Conversation Coordinator* is devoted to control the cooperation. It defines one or more policies of cooperation that can be followed during the execution of the application, and the roles that can be covered by the users of the application. The Conversation Coordinator receives the requests of role change by the Tools through a bi-directional control communication channel, and manages them according to the policy of cooperation that is currently in place. It also controls the Communication Coordinators by sending them, through unidirectional command communication channels, the information about the access rights of the users accessing the shared workspace. The Communication Coordinators forward the actions performed by a user on a tool according to these access rights, ignoring the actions if the user has not the input access right, and sending them to all the users having the output access right in the other case.

The hierarchy of Coordinators that has to be executed to control the cooperative application is dynamically defined at start-up. This means that, for example, it is possible to develop a catalogue of Conversation Coordinators for an application, each one defining different policies of cooperation, and to choose the right one at start-up time.

The *Supervisor* provides services that make it possible to start and manage several cooperative applications. In general, initiating a cooperation is a complex procedure, that can be executed according to some policy, and can involve many people. In fact, the participants together with their initial roles have to be selected, the policy of cooperation and the time-line of the cooperation have to be set-up, the participants have to be allowed to know that the cooperation is going to be started, finally, a cooperative application able to support the cooperation has to be chosen. ImagineDesk does not supply a specific support for these issues. Instead, it supplies, through the Supervisor, some basic services that can be composed to carry out this task.

4. Integration Strategies

SPADE and ImagineDesk offer complementary functionalities that can be jointly exploited to offer more effective process support to software developers. In particular, SPADE can offer explicit process modeling and enactment, tool integration facilities, and product modeling and management of artifacts. ImagineDesk offers support to build and execute multimedia, cooperative applications. It supplies a powerful compositional approach to build cooperative applications, but do not allow dynamic definition of the policies of cooperation.

We have identified different integration strategies, that try to take advantage of the different features offered by the two environments. In particular, two extreme strategies are presently being prototyped and evaluated: loose integration, and process coupling.

Loose Integration

In this case, in SLANG we model the procedures used to start the cooperation. Basically, SPADE launches and controls the ImagineDesk Supervisor (using SLANG constructs to invoke external tools), but does not interfere with the evolution of the cooperative application (see Figure 3). This means that the cooperative application is seen by SPADE as any other software development tool that can be launched and controlled through appropriate SLANG process model fragments. The policies used to control and run the cooperative application, therefore, are kept into ImagineDesk Coordinators, and are distinct and not integrated with the software development policies.

The advantage of this solution is obviously simplicity. The integration is achieved in a very straightforward way, by simply modeling in SLANG the invocation of the Supervisor services that allow the cooperation to be organized and the cooperative application to be started. Notice that these phases are modeled as any other software process procedure.

On the other hand, it is impossible to integrate from a process viewpoint the procedures described in both the ImagineDesk Conversation Coordinators and in the SLANG process model. Actually, a Conversation Coordinator implements a simple process,

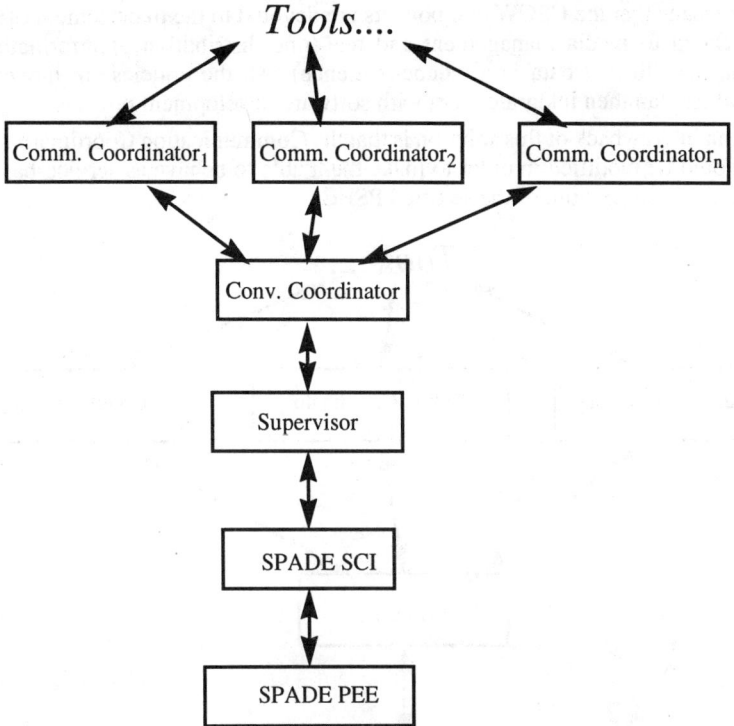

Fig. 3. Loose integration of SPADE and ImagineDesk.

whose semantics is hard-coded and defined by the Conversation Coordinator implementation. In many cases, it would be important to "blend" and integrate the two processes in order to reduce redundancy and increase effectiveness. For instance, a Conversation Coordinator includes information on users and their rights. Similar information are usually stored in the software process model. Similarly, decisions related to the invocation of specific conversation operations might depend on the state of the software process being executed (and vice-versa). Moreover, if the policies of cooperation are kept into ImagineDesk Conversation Coordinators, it is not possible to take advantages of SLANG reflective features to dynamically changes processes. Right now, changing a Conversation Coordinator policy requires a change in its coding, and cannot obviously be accomplished on-line.

Process Coupling

Clearly, a more effective solution consists of the direct integration of the Communication Coordinators in SPADE (see Figure 4). In SPADE, specific process model fragments are defined to control Communication Coordinators and Tools.

This solution basically extracts all the policies from ImagineDesk and makes them explicit as SLANG process model fragments. These model fragments replace the Conversation Coordinators and the Supervisor, and model: 1. the cooperative applications the start-up; 2. the policy of cooperation, and the management of the requests of role change coming from the Tools; 3. the way the underlying Communication Coordinators are controlled.

This means that the CSCW components are devoted to the management of specific issues related to media management and real-time distribution of information (i.e., managing the flow of data in a videoconference). All the policies are moved to the PSEE, which can then integrate them with software development policies.

The main drawback of this solution is that the Communication Coordinators and the Tools should be modified in order to make them able to behave as service-based tools of SPADE (or, in general, of the selected PSEE).

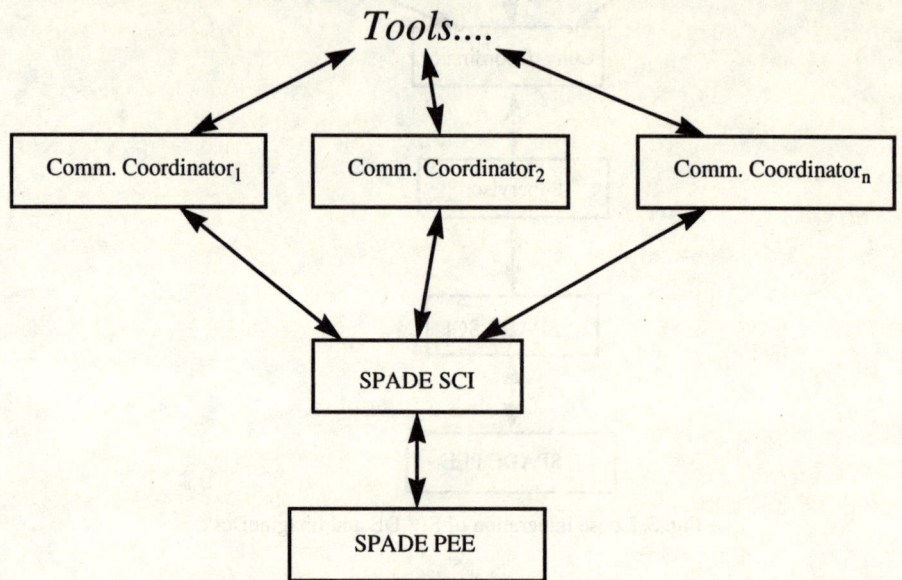

Fig. 4. Process coupling of SPADE and ImagineDesk.

5. Conclusions

This paper has briefly analyzed an important issue that is currently being addressed by several researchers (see for example [2]). Namely, we have sketched some problems related to the integration of CSCW and PSEE technologies, by analysing the characteristics of two research prototypes: SPADE and ImagineDesk. We have briefly discussed two solutions that we have envisaged. The first solution keeps the two environments logically distinct. ImagineDesk is simply launched and controlled at a very coarse-grain level by SPADE. In the second solution, SPADE is used to model the policies of cooperation that are presently implemented in ImagineDesk.

We believe that the two alternatives discussed in the previous sections can be fruitfully used to identify more general requirements and issues in integrating CSCW and PSEE technologies. From our experience and from the assessment of other existing prototypes and products, we argue that most CSCW environments include process-specific parts. In most cases, these processes are hard coded or specified using quite primitive languages. Conversely, PSEEs do not take into account the issue of synchronous, multimedia cooperation. Our solution based on process coupling aims at extracting all process knowledge from the CSCW environment and at modeling it as part of the software process model.

Notice that this approach is quite similar to what it has been occurring to software development tools. With the introduction of tool integration facilities such as Field (and its commercial spin-offs Tooltalk, FUSE, and BMS), it is now possible to extract from software development tools the knowledge about the policy of interaction with other tools, and store it in the message server. Even more, several PSEEs are taking advantage of this architecture to better integrate and control tools at a very detailed level (see for instance ProcessWeaver, SPADE, and EPOS). In this paper, a similar approach is advocated as far as the integration of CSCW and PSEE technologies is concerned. We believe that by clearly distinguishing the roles, advantages, and characteristics of this two technologies we can greatly improve the effectiveness of environments supporting software development activities.

6. Acknowledgments

The SPADE and ImagineDesk projects are carried out at CEFRIEL and at Politecnico di Milano. We wish to thank our colleagues who are contributing to these projects, and, in particular, Sergio Bandinelli, Flavio De Paoli, Luigi Lavazza, and Silvano Pozzi.

References

1. P. Dewan, B. Krishnamurthy. Relations between CSCW and Software Process Research: a position statement. To appear in the Proceedings of the 9th International Software Process Workshop. Airlie (Virginia), October 1994.
2. I. Ben-Shaul, T. Heineman, S. Popovich, P. Skopp, A. Tong, G. Valetto. Integrating Groupware and Process Technologies in the Oz environment. To appear in the Proceedings of the 9th International Software Process Workshop. Airlie (Virginia), October 1994.
3. S. Bandinelli, A. Fuggetta, C. Ghezzi, L. Lavazza. SPADE: an environment for software process analysis, design, and enactment. In A. Finkelstein, J. Kramer, and B. Nuseibeh, eds. Software Process Modeling and Technology. Research Studies Press (J. Wiley), 1994.
4. S. Pozzi, E. Di Nitto. ImagineDesk: a Software Platform Supporting Cooperative Applications. In the Proceedings of ACM 1994 Computer Science Conference (CSC94). Phoenix (AZ), Mach 1994.
5. S. Bandinelli, M. Braga, A. Fuggetta, L. Lavazza. The architecture of the SPADE-1 PSEE. In Proceedings of the Third European Workshop on Software Process Technology (EWSPT 94). Springer-Verlag, LNCS Series, 1994.

Distributed Modelling Session

Gregor Engels

Leiden University, The Netherlands

A fine-grained modelling of all aspects of a software process inherently results in large and complex specifications. In order to overcome this complexity, software process models have to be appropriately modularized. This means that the specification is spread over several, interrelated subspecifications, where each subspecification describes a certain view on the software process.

The aim of this session is to discuss this topic in more detail. In particular the following questions are addressed:

1. Which views on a software process have to be modelled?
2. Does a view describe a certain aspect of a software process (e.g. the product view) or does a view describe a complete software process from a certain point of view (e.g. the manager view).
3. What are appropriate specification means to describe a certain view?
4. What are the interrelations between different views and how can they be integrated into a comprehensive software process model.
5. How can distributed, multi-view specifications be enacted?
6. What are the benefits and drawbacks of distributed, multi-view process modelling with respect to understandability, maintainability, changeability, and reusability of software process models?
7. What can be learnt from other research areas and what are open research issues within the software process modelling area?

The selected papers in this session discuss answers to these questions and form a well-suited starting point for intensive and active workshop discussion.

The long paper by Graw and Gruhn discusses means for a distributed modelling of several, interconnected process models. The position papers by Estublier, by Nuseibeh et al. as well as by Groenewegen and Engels discuss the notion of a view. The position paper by Harmsen and Brinkkemper describes a language to compose so-called method fragments, which may also be applied to compose process views.

Process Management In-the-Many

Günter Graw

Volker Gruhn

LION GmbH

Universitätsstraße 140

44799 Bochum

Germany

Abstract. Software process management and business process management are areas of research for some years now. Both suffer from considering processes as isolated entities. In contrast to that, we believe that software processes and business processes are impacted by many other processes surrounding them. In this article we motivate why we understand processes as entities which communicate with each other. We suggest mechanisms to support modeling and enaction of communicating processes.

Keywords

process management, process enaction, enaction in-the-many,
communicating processes

1 Introduction

Software processes are subject of research for nearly a decade now. Many prototypes have been built and even some products have been developed [BBFL94, DG94, PS92, EG94]. Most of them allow software process modeling, some of them support software process analysis, and a few actually support software process enaction. In the software process community the question of what a software process modeling should like is discussed intensively [O'C93, War94].

Business processes are dealt with in industrial practice for some years. Modeling of business processes is considered as prerequisite for process enaction, but not as a virtue in itself. Most business process modeling languages offered in commercial workflow management systems are not very comfortable [HL91]. This makes it difficult to adapt process models to changing conditions.

In [DK76] the notion of *programming in-the-large* is discussed. It is pointed out that development of large software systems with large teams looks different from programming small systems in small teams. The same applies to the management of software processes and business processes [BFG93a].

Both suffer from missing support for

- simultaneous modeling of related processes,
- analysis, which focuses on properties not being local to exactly one process, but to networks of related processes,

– simultaneous enaction of related processes and communication between them.

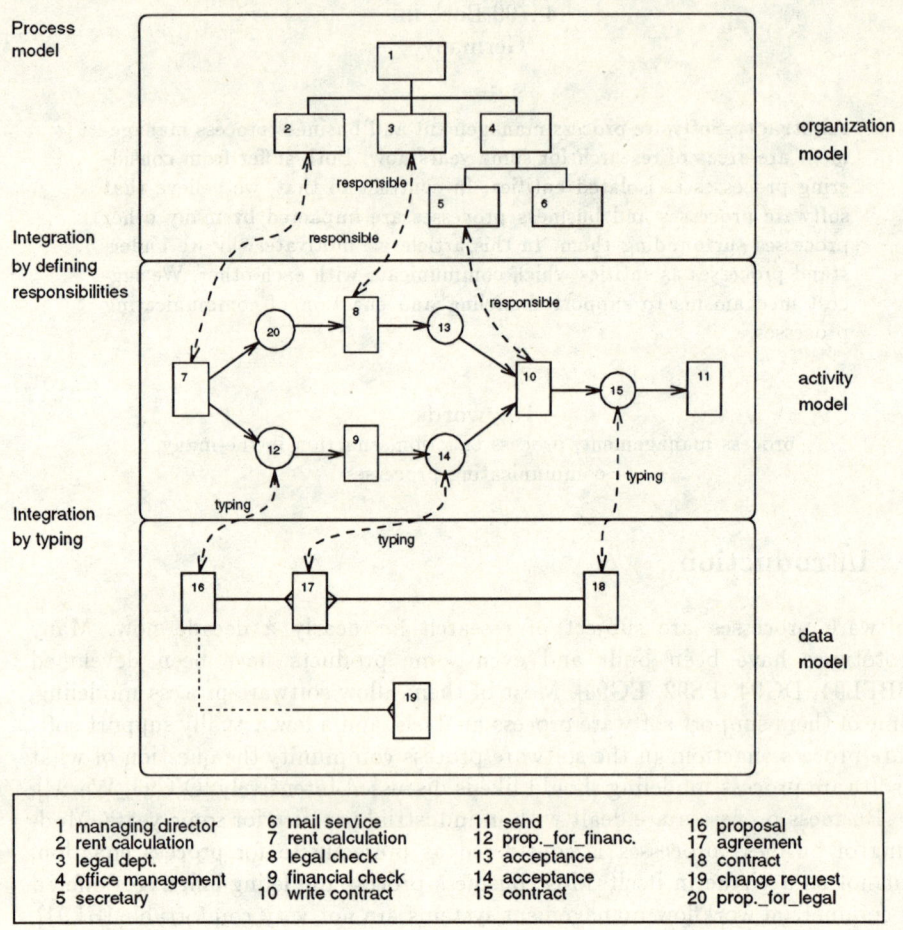

Process model				organization model
Integration by defining responsibilities				activity model
Integration by typing				data model

1 managing director	6 mail service	11 send	16 proposal
2 rent calculation	7 rent calculation	12 prop._for_financ.	17 agreement
3 legal dept.	8 legal check	13 acceptance	18 contract
4 office management	9 financial check	14 acceptance	19 change request
5 secretary	10 write contract	15 contract	20 prop._for_legal

Fig. 1. Integration of data, activity, and organization models

Process management which overcomes these shortcomings is called **process management in-the-many** in the following. We identified the need to manage many processes in a business process project we have been involved in recently. In this project (called WIS) the task was to develop a system supporting all processes of housing construction and administration. These processes included, for example, building of apartments, repair management, financial management, accounting, and real estate management. Since most of these processes have important interactive parts and since most are well-understood and structured, we decided to develop the WIS system in a process management approach. In other words, the processes for housing construction and administration were modeled, analyzed and enacted. Since there was a broad range of processes to be managed it was inevitable to involve many modelers, because this was the only way to acquire all the process knowledge needed. In fact, about 60 persons were involved in process modeling, 20 of which were directly responsible for particular processes, while the others were consultants. Thus, it was necessary to let many persons model *their* processes simultaneously. Moreover, we had to consider processes as entities, between which interfaces had to be agreed (the process for financial administration of repairs, for example, has to exchange information with the accounting process). During process enaction, finally, it was necessary to enact many processes at once. A process for repair management, for example, can be active while processes for real estate management and accounting are active as well. Enaction of many processes at once was also necessary for processes following one model. Several processes for rent increase may be active at once, if they, for example, process rents of apartments in different regions.

Commercially available process management environments did not fulfill our ideas of simultaneous process modeling, interface definitions between processes and/or integration between data models, activity models and organization models. Thus, for the development of the WIS system, the LION engineering environment LEU [Gru94, DGZ94] was developed. LEU implements the FUNSOFT net approach to process management [DG90]. It is based on data modeling, activity modeling and organization modeling:

- **Data models** are used to describe the structure of objects (and their relationships), which are manipulated within a process. In the FUNSOFT net approach data models are described by means of extended entity/relationship diagrams [Bar90]. A scheme of a data model is shown in the bottom part of figure 1. Object types like *change requests, proposals, contracts* are described as entity types. Object types can either be of a predefined format (postscript, WordPerfect, etc.) or they can be structured. In case of structured object types attributes have to be defined. A *contract*, for example, could be defined by a contract number of attribute type *integer*, by the contract parties identified by name (and, therefore, of type *string*), and by a contract text of type *text*. Relationships between object types can be of different cardinalities. They can be optional or mandatory. The relationship between *proposals* and *change requests*, for example, is of cardinality 1:n, which means that one *proposal* can be related to several *change requests*, but that each *change request* can be related to only one *proposal*. Moreover, this relationship is optional:mandatory, which means that a *proposal* need not be related to a *change request*, but that each *change request* must be related to a *proposal*.

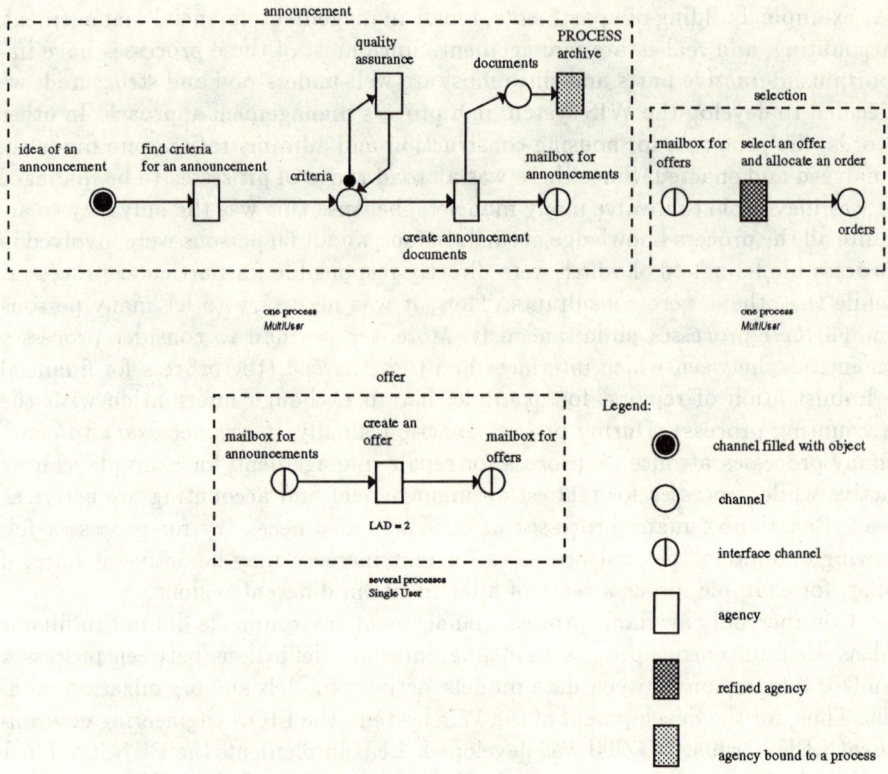

Fig. 2. Relationship between process models of the scenario

- **Activity models** are used to define activities to be executed in a process. Activity models are described by FUNSOFT nets [EG91]. FUNSOFT nets are high level Petri nets [Rei86], whose semantics is defined in terms of Predicate/Transition nets [Gen87]. A scheme of a FUNSOFT net is sketched in the central part of figure 1. Activities to be carried out in processes are described by agencies (represented as rectangles). Agencies in FUNSOFT nets can be fired (i.e. the corresponding activity can be executed) as soon as all required input objects are available. Objects produced and manipulated by activities are stored in channels (represented as circles). FUNSOFT nets do not only contain a definition of activities and their parameterization, but also an order of activities. They allow to define that activities have to be carried out sequentially, concurrently or alternatively. The order is based on preconditions which have to be fulfilled before an activity can be executed. Activity *write contract*, for example, can only be carried out, when an *acceptance* acknowledgement is produced by the legal and by the financial check.

- **Organization models** are used to define which organizational entities are involved in a process. The top part of figure 1 sketches an organization model described by organization diagrams. It identifies, for example, that the top level organizational entity is that of the *managing director*. Subordinated to that entity, we find the organizational entities *rent calculation, legal dept.* and *office management*. These entities are related to roles. Roles are sets of permissions for the execution of activities and for the manipulation of objects of certain types. A person assigned to an organizational entity plays the roles of that entity. The relationship between organizational entities, roles and persons is defined in a tabular form.

Once these aspects of processes have been modeled, it is necessary to integrate them. Integration of data models, activity models, and organization models means to define:

- which channels of FUNSOFT nets are typed by which object types identified in data models (indicated by the arrows annotated with *typing* in figure 1); this corresponds to definition and use of object types as defined in SPADE [BFG93a],
- which organizational entities are responsible for which activities (indicated by the arrows annotated with *responsible* in figure 1).

Based on the experience gained in the WIS project we identified some requirements for process management in-the-many. These are discussed in section 2. Section 3 discusses an example of related processes and discusses the LEU support for process modeling in-the-many. Then, section 4 sketches what process enaction of many processes looks like. Finally, section 5 concludes this article pointing out which further in-the-many support is needed.

2 Requirements for Process Management in-the-Many

In the following we discuss why process management in-the-many is needed in order to succeed in transferring process technology into practice and which requirements for process management in-the-many exist.

Networks of Processes are Real Life Processes in general are embedded into surroundings in which other processes run. A software processes for the development of a certain piece of software is embedded into and directly influenced by processes of hiring personnel (because the availability of qualified staff is a key factor for successful software development), of selling other products (because of budgetary reasons) and by general management processes (because management decisions may affect technical issues). Thus, realistic process models should reflect that processes communicate which other. This is particularly important in decentralized organizations where

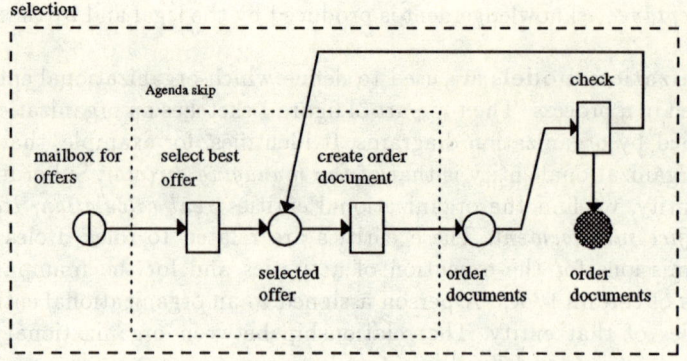

Fig. 3. Refinement of select an offer an allocate an order

more or less autonomous divisions and departments are responsible for *their* part of the process.

Simultaneous modeling For related processes it is necessary to allow that they are modeled simultaneously. This is due in order to keep process complexity manageable and it is also necessary in order to involve modelers who have the required process knowledge.

Modeling different types of communication between processes Related processes communicate with each other. We have to support different types of communication. The easiest communication is between one defined sender process and one defined recipient process (one sender, one statically defined recipient). Sometimes, the recipient is not defined at modeling time, but it is selected while the sending process is running (one sender, one dynamically defined recipient). Moreover, there are two types of communication between one sender process and several recipient processes. Again we have to distinguish between static and dynamic definition of the multiple recipients (one sender, multiple (statically or dynamically) recipients).

Embedding process models into other process models Certain processes are used within different contexts. Examples of such processes are processes for archiving documents, reviewing documents and arranging meetings. Processes like these are not specific for a certain application domain, but they are needed in various situations. In order to ensure, that they are not modeled individually in each situation, it is necessary to administrate a set of basic process models which can be embedded into other processes. This kind of reuse of process fragments was already suggested in [BFG93a].

To reuse process models it is necessary to embed a process (e.g. an *archive* process into another process which produces documents). During modeling it is necessary to model how such a process is embedded and how its input and output behavior is integrated into the embedding process.

New types of analysis Traditional process analysis focuses on properties which are local to individual processes. If we consider related process models, then process analysis has to be extended to properties of process model networks. Examples of such properties, are the number of interfaces between processes, the types of objects which are exchanged between processes and similar properties which determine process networks. Moreover, analysis of delays which occur because one process waits for results of another process and investigations of process parallelism are worthwhile to identify where process networks can be speeded up.

Sophisticated process monitoring and tracing Process monitoring and tracing for related processes is more difficult than for single and isolated processes. In addition, it has to be recorded which objects are exchanged between which processes, when which processes are started, interrupted, resumed, and terminated. Moreover, it is worth to record which processes are waiting for objects produced by which other processes.

Scalability Processes are very often enacted within dynamic organizations. Thus, it is important to provide the opportunity to adapt firstly the process models themselves, but secondly also the number of processes running. A growing company easily will run into the situation to deal with a growing number of projects. Thus, it is necessary to enact multiple processes in parallel.

Different types of process modification If several processes following one model are enacted in parallel, it can become necessary to apply modifications either to only a few processes or to all processes following that model. If, for example, an error is detected in process model M while processes $p_1, \ldots, p_n$ following M are active, then a correction has to be applied to all processes $p_1, \ldots, p_n$. This corresponds to a model modification which is propagated to all processes following that model. If on the other hand, only process p_1 has run into a process-specific problem, then a corrective action only has to be applied to p_1. For a more detailed classification of types of process model modifications we refer to [BFG93b].

Process administration Understanding systems as communicating processes means to handle processes as first class objects. Accordingly, operations to manipulate processes are needed. These operations have to support start of processes, interruption of processes, resume of processes, and termination of processes.

In the following sections we focus on modeling in-the-many and enaction in-the-many as implemented in LEU. Analysis mechanisms are discussed in more detail in [BG93].

3   Process Modeling in-the-Many in LEU

In order to explain the modeling in-the-many features as implemented in LEU we introduce a small example at first. This example covers the process of assembling an announcement, sending it to all potential suppliers, await their offers, choose the best and, finally, giving the order to exactly one supplier. This context is represented by several related process models. Figure 2 shows that four process models are involved. The structure of three of these four process models is shown in figure 2 explicitly (*announcement, selection, offer*). The fourth process model (*archive*) appears as annotation to one agency in the *announcement* process model. It is not discussed in detail. One thing worth to mention is that agency *quality assurance* accesses channel *criteria* by a COPY-edge (graphically represented by a circle at the edge's foot). This means, that it reads objects from channel *criteria*, but that the read object is not removed from the channel. Process model *selection* contains a refined agency, called *select an offer and allocate an order*. This means, that details of the *selection* process are not shown on the top level. Details of this activity are described by the FUNSOFT net shown in figure 3.

Details of the process models are discussed in the following explanation of in-the-many features.

- We can distinguish process models for which it is crucial that there exists only one process following that model at any time and others for which several processes can be active. If we revert to our example, it is obviously possible that several *offer* processes can be enacted simultaneously. On the other hand, there should only be one *selection* process at any time in order to ensure that there is an unambiguous order at the end. Thus, for each process model it has to be modeled, whether or not more than one process can be enacted simultaneously. In figure 2 we recognize a corresponding definition at the bottom of the process boxes. There can be several processes following model *offer*, but only one process of model *selection* and *announcement*.
- Processes can be closely linked together by binding process models to agencies of other processes. To bind a process model M_2 to an agency t of a process model M_1 means:
  - A process m_2 following M_2 is started as soon as objects are written to channels which are read by agency t. Process m_2 is called like a procedure. Its results are returned to the calling process. In detail, the results are written to the channels to which agency t writes. Thus, a synchronous communication between calling and called process is established.

    An example is the grey agency of process model *announcement* (compare figure 2). To this agency the process model *archive* is bound. As soon as an object arrives in channel *documents* an *archive* process is started.
  - As soon as no activity in the started process can be carried out (in FUNSOFT net terminology: the net is dead), the process is terminated.

  The modeling construct of binding processes to agencies supports reuse of process and it supports modularization of complex processes. This notion of binding process models to agencies and the way to execute them corresponds to the notion of activity execution proposed in [BFG93a].
- LEU supports parallel work of different users on the same activity. The laddering attribute is introduced to model the maximum number of simultaneous executions of an activity. A laddering attribute of 1 means that the

activity cannot be executed simultaneously to itself. The agency *create an offer* of the process model *offer* shown in figure 2 has a laddering attribute with the value 2. This means that at maximum 2 offer creation activities might be carried out simultaneously. Laddering attributes can also be defined for agencies to which processes are bound. In the example discussed the agency to which the process *archive* is bound has a laddering attribute of 1, which is the default value and which is, therefore, not explicitly mentioned in figure 2.

- Processes following different process models can exchange information via so-called interface channels. If a process m_1 following a model M_1 writes an object o_1 to an interface channel s_1, then o_1 is sent to all processes in whose models an access to s_1 is modeled. In detail, interface channels support the four types of communication identified above:

    • The communication between two processes is established in an asynchronous mode. The channel *mailbox for announcements* in figure 2 is an interface channel. It appears in the process models *announcement* and *offer* (one sender, one statically defined recipient).

    • Interface channels can also be used to enable the communication between more than two processes. By writing an object into an interface channel, the sending process communicates with all processes in whose models the interface channel occurs. The object is sent to all recipients (one sender, multiple statically defined recipients).

    • If one or multiple recipients have to be defined dynamically (i.e. at enaction time), the object which has to be sent is tagged by the names of one (one sender, one dynamically defined recipient) or several (one sender, multiple dynamically defined recipient) recipients. Then the object is sent only to the identified recipients.

4 Process Enaction in-the-Many in LEU

According to modeling, LEU supports the enaction of many processes. Thus, we discuss the enaction components and mechanisms provided for enaction in-the-many.

- Agendas are the LEU means to offer activities to be carried out in different processes to process participants. Activities to be carried out are displayed to process participants by means of personal agendas. The agenda of a process participant contains at any moment all activities he is allowed to carry out. Only activities which demand interaction (manual activities for short) appear in agendas, automatic activities are not passed to agendas, but executed automatically as soon as all their input objects are available The execution of a manual activity can, for example, mean to call an office automation tool or to start a dialogue in which data has to be manipulated. After an activity has terminated its entry is deleted from all agendas.

    Depending on the number of users which are allowed to participate in a process, a process model can be defined as single-user or multiple-user. In a multi-user process many users are allowed to participate in different activities. This supports the cooperative work of team members within one process. While process models usually describe processes in which several process participants are involved, there are some situations where it is useful to model that certain processes or process parts should be done by just one participant. This is, for example, the case for the activities *select best offer* and *create order document* of the example discussed. Even though there may be many potential participants who may select offers, it is useful to define that the person who selected the offer shall also create the order documents. If a process model is defined as single-user process model, then in each process following that model only one person participates. Once a single-user process (a process following a single-user process model) is started, the first activity to be executed is offered to all potential participants. When one of them decides to participate, then this process is completely delegated to that participant. All further activities are only offered to this person.

For certain parts of a process which are to be carried out by one participant it is possible to define, that activities should not appear in agendas, but that they are executed immediately. This can be useful, if certain strongly related activities are to be executed subsequently. In the example such an *agenda skip* is defined for the activities *select best offer* and *create order document*. Thus, the process participant who executes activity *select best offer* has to execute *create order document* as soon as *select best offer* is finished.

– During enaction all modeling in-the-many features introduced above (several processes following one model, process models bound to agencies, laddering of agencies and processes, interface channels and different types of communication) are supported. The start and termination of processes is done automatically where necessary. Information between communicating processes is transported automatically. This will be illustrated by the scenario discussed at the end of this section.

– Even though some processes can be started and terminated automatically, it remains necessary to define start and termination of some others manually. If we look at the *offer* process, for example, then this process either can be started at the very beginning (in this case it waits for announcements) or it can be started as soon as an announcement arrives in channel *mailbox for announcements*. If it is supposed to be running from the very beginning, someone (usually a system supervisor) decides about when to start an *offer* process. This supervisor also needs support for interrupting a particular process, all processes following a certain model or even all processes (e.g. when the underlying model has to be modified or when a hardware exchange is necessary), for resuming processes and for manually terminating processes. These operations are implemented in the LEU controller.

Figure 4 shows the relationship between different enaction components and their embedding into the overall architecture. Boxes represent components (i.e. module hierarchies). Arrows between boxes indicate that there is a *call-*relationship between these components. The annotations of arrows give examples of demanded services. The component *Process engine*, for example, calls the service *start activity* from component *Automatic activity handler*. Between some components we find double-headed arrows (e.g. between *Process engine* and *Agenda controller*). In that case the arrow annotations indicate which service is demanded by which component (e.g. the *Process engine* demands to *fill entries* into personal agendas, and the *Agenda controller* returns entries which have been selected (service *selected entry*)). The main functionality of these components and their interfaces to other components are discussed in the following:

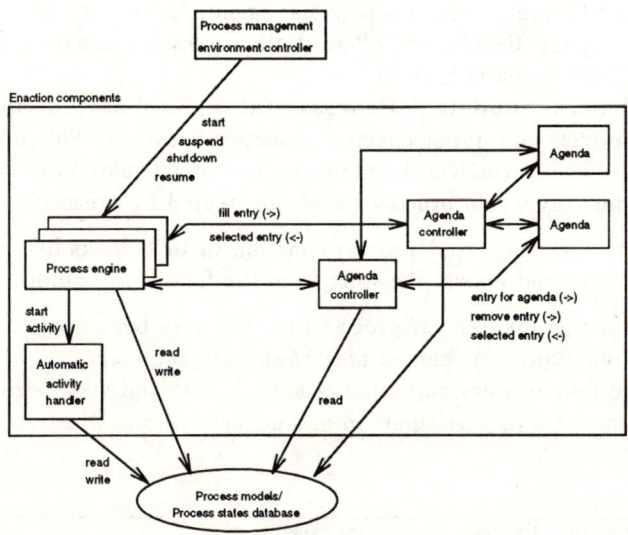

Fig. 4. Potential architecture of enaction components

- There is one **Process engine** for each process running. The process engine for a process P identifies all activities of P for which all input objects are available and which, therefore, could be executed. In case of automatic activities, it starts them via the *Automatic activity handler* (compare below). In case of manual activities, corresponding entries are sent to the process engine's *Activity controller*. To identify executable activities, a process engine reads the database storing process models and process states. To update the process state after executing an activity, the process state database is modified.
- The only functionality of the **Automatic activity handler** is to receive execution requests from all process engines, to start the requested activities, and to return whether or not the execution has been successful.
- As soon as a process engine is started, a related **Agenda controller** is created. An agenda controller administrates all manual activities of the related

process. It is connected to the personal agendas of all process participants who may participate in the process. The process participants who may participate can be identified on the basis of roles and permissions which are also stored in the underlying database. Agenda controllers manage manual activities. If a manual activity could be executed, the agenda controller sends it to the personal agendas of all process participants who could participate. If an activity is selected from an agenda by a process participant, the agenda controller returns this information to its process engine which starts the activity at the workstation of the corresponding participant.

- There is one personal **Agenda** for each process participant. As soon as a participant logs into LEU, his personal agenda is started and automatically connected to the agenda controllers of all processes, the participant has permissions/roles to participate in.
- In the **Process models / Process states database** information about process models and about enacted processes is stored. This information is accessed to check who has the permissions to participate in activities and in order to identify which process states are created by executing activities.

In order to illustrate what process enaction in LEU looks like, we revert to the example discussed above. We start from the following assumptions:

1. There is one *announcement* process from the very beginning.
2. There is one object in channel *idea of an announcement*.
3. There are four process participants logged in at the very beginning. They have permissions to participate in manual activities as illustrated in table 1.

Person	Process	Activities
Miller	announcement	find criteria for an announcement
		create announcement documents
	archive	
Smith	announcement	quality assurance
	offer	check
	archive	
Johnson	selection	select an offer and allocate an order
Lansing	offer	create an offer

Table 1. Persons and their responsibilities

Based on these assumptions we discuss an enaction scenario which firstly reflects the view of a process participant (by means of the agendas) and which secondly gives an insight into the interaction LEU of enaction components.

Figure 5 represents an overview of the enaction states and transitions between them in the form of snapshots. Each state is composed of three entries which are discussed in the following:

1. The agendas are represented as boxes. As top annotation of agendas we recognize the names of process participants. Moreover, the agendas contain the names of activities to be carried out.
2. The agenda controllers are represented by ellipses. Each of them carries a name composed of the prefix AC which is concatenated with the number of the current process.
3. The process controllers are represented by ellipses. Each of them carries the name of its associated process model.

The overall state of processes and agendas is modified when an activity is carried out by a process participant. The names of the executed activities causing the transitions are given at the bottom of each state.

The process participants Miller and Smith have the permission to execute the activities of the *announcement* process in *state1* of the scenario. Thus, their agendas are connected to the agenda controller AC1 which is connected to the process engine of this process model. The activity *find criteria* can be executed because of the initial object in channel *idea of an announcement* (compare figure 2) and is displayed in Miller's agenda. Miller executes the activity *find criteria* which causes a transition to *state2*. In *state2* the activities *create announcement documents* and *quality assurance* are enabled. Smith executes activity *quality assurance* and the state changes. The states *state2* and *state3* are similar because activity *quality assurance* accesses its input object by a COPY-edge. Miller executes the activity *create announcement documents* in *state3*. Activity *create announcement documents* writes an object into interface channel *mailbox for offers* and into channel *documents*. This is the reason for two events in *state4*. Firstly, a process engine for the process *offer* and its associated agenda controller AC2 are started and connected to Lansing's agenda. Secondly, a new process engine for the process *archive* is started together with its associated agenda controller AC3. This agenda controller connects to Miller's and Smith's agendas. Johnson logs out in this state and his agenda disappears. Lansing executes the activity *create an offer* which writes an object to the interface channel *mailbox for offers* and the process engine for the *selection* and the agenda controller A4 are started. The agenda controller A4 connects to Lansing's agenda. Lansing executes the *archive* process which causes a transition to *state6*. The agendas of Miller and Smith are updated correspondingly. In this state the process engine of the *archive* process is terminated and the process engine and its associated agenda controller AC3 are terminated. The FUNSOFT net of the *announcement* process is dead in this state. Thus, the process engine and its associated agenda controller AC1 are terminated.

5   Conclusion

In this article we motivated the need to understand processes as inter-dependent. We motivated what is required to model their relationships and we discussed how

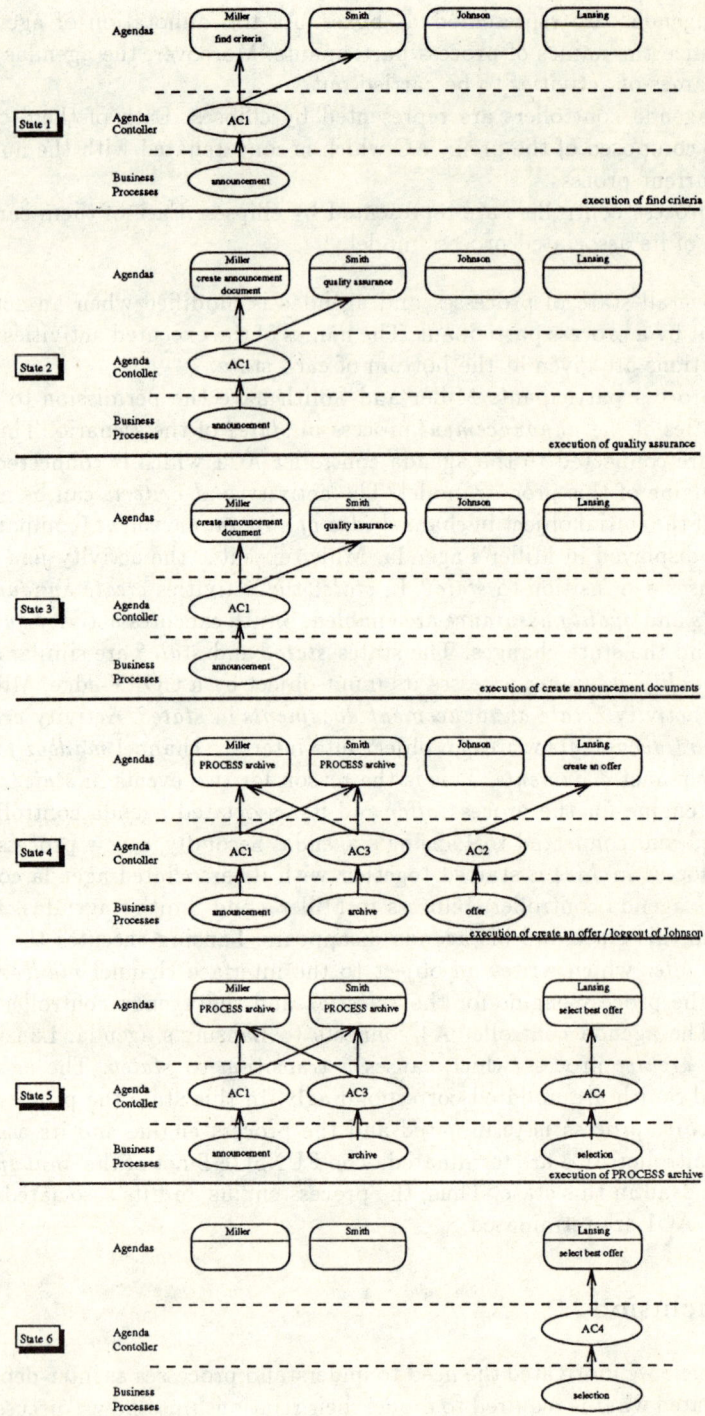

Fig. 5. Enaction scenario

their relationships impact process enaction. The mechanisms suggested for process management in-the-many have been implemented and were used in recent business process management projects.

Our experience in the area of housing construction and administration processes, insurance processes, and customer care processes is that it is inevitably necessary to manage several processes at once. This helps to reflect organizational issues, responsibilities of process participants for certain objects and to adapt process organization to changing requirements and circumstances. Altogether, we believe that proper structuring of complex process models into manageable parts was a key success factor in the process management projects recently carried out.

Further research is necessary to identify how performance requirements for time-critical processes can be met and how process engines can be distributed to clusters of workstations (or even other machines). Another open issue is the question how process management can be introduced without demanding to develop new software systems from scratch. In most cases it will be necessary to just model a few processes and to let them communicate with *old* software. This will lead to a variety of integration problems between process management systems on the one hand and legacy systems and standard software on the other hand.

References

[Bar90]   R. Barker. *CASE*Method Entity Relationship Modelling*. Addison-Wesley, Wokingham, England, 1990.

[BBFL94] S. Bandinelli, M. Braga, A. Fugetta, and L. Lavazza. *The Architecture of SPADE-1 Process-Centered SEE*. In B. Warboys, editor, *Software Process Technology - Proceedings of the 3rd European Software Process Modeling Workshop*, pages 15–30, Villard de Lans, France, February 1994. Springer. Appeared as Lecture Notes in Computer Science 772.

[BFG93a] S. Bandinelli, A. Fugetta, and S. Grigolli. *Process Modelling In-the-Large with SLANG*. In *Proceedings of the 2nd International Conference on the Software Process - Continuous Software Process Improvement*, pages 75–83, Berlin, Germany, February 1993.

[BFG93b] S.C. Bandinelli, A. Fuggetta, and C. Ghezzi. *Software Process Model Evolution in the SPADE Environment*. IEEE Transactions on Software Engineering, 19(12), December 1993.

[BG93]   A. Broeckers and V. Gruhn. *Computer-Aided Verification of Software Process Model Properties*. In *Proceedings of the Fifth Conference on Advanced information Systems Engineering (CAiSE)*, Paris, France, June 1993.

[DG90]   W. Deiters and V. Gruhn. *Managing Software Processes in MELMAC*. In *Proceedings of the Fourth ACM SIGSOFT Symposium on Software Development Environments*, pages 193–205, Irvine, California, USA, December 1990.

[DG94]   J.-C. Derniame and V. Gruhn. *Development of Process-Centered IPSEs in the ALF Project. Journal of Systems Integration*, 4(2):127–150, 1994.

[DGZ94] G. Dinkhoff, V. Gruhn, and M. Zielonka. *Praxisorientierte Aspekte der LEU-Datenmodellierung (in German)*. *EMISA Forum*, (1), January 1994.

[DK76] F. DeRemer and H.H. Kron. *Programming-in-the-Large versus Programming-in-the-Small*. *IEEE Transactions on Software Engineering*, 2(2), February 1976.

[EG91] W. Emmerich and V. Gruhn. *FUNSOFT Nets: A Petri-Net based Software Process Modeling Language*. In *Proc. of the 6th International Workshop on Software Specification and Design*, Como, Italy, September 1991.

[EG94] G. Engels and L. Groenewegen. *Specification of Coordinated Behavior by SOCCA*. In B. Warboys, editor, *Software Process Technology - Proceedings of the 3rd European Software Process Modeling Workshop*, pages 128–151, Villard de Lans, France, February 1994. Springer. Appeared as Lecture Notes in Computer Science 772.

[Gen87] H.J. Genrich. *Predicate/Transition Nets*. In W. Brauer, W. Reisig, and G. Rozenberg, editors, *Petri Nets: Central Models and Their Properties*, pages 208–247, Berlin, FRG, 1987. Springer. Appeared in Lecture Notes on Computer Science 254.

[Gru94] V. Gruhn. *Communication Support in the Workflow Management Environment LEU*. In *Connectivity '94 - Workflow Management - Challenges, Paradigms and Products*, pages 187–200, Linz, Austria, October 1994. R. Oldenbourg, Vienna, Munich.

[HL91] K. Hales and M. Lavery. *Workflow Management Software: the Business Opportunity*. Ovum Ltd., London, UK, 1991.

[O'C93] L. O'Conner, editor. *Proceedings of the 2nd International Conference on the Software Process - Continuous Software Process Improvement*, Berlin, Germany, February 1993.

[PS92] B. Peuschel and W. Schäfer. *Concepts and Implementation of a Rule-based Process Engine*. In *Proceedings of the 14th International Conference on Software Engineering*, Melbourne, Australia, May 1992.

[Rei86] W. Reisig. *Petrinetze (in German)*. Springer, Berlin, FRG, 1986.

[War94] B. Warboys, editor. *Proceedings of the 3rd European Workshop on Software Process Modelling*, Villard de Lans, France, February 1994. Springer. Appeared as Lecture Notes in Computer Science 772.

A Generalized Multi-View Approach

Jacky Estublier and Noureddine Belkhatir
L.G.I. BP 53X 38041 Grenoble
FRANCE

Abstract. It is advocated here that integrating abstraction and modularity into the concept of point of view, and extending the view concept to the process itself (and not only to data used by processes), provides an uniform conceptual framework for aspects like agent point of view, quality models and process monitoring..

1 Introduction

Within Process Modelling, formalisms have been proposed to satisfy requirements like modularity and , abstraction. On the other hand, formalism have been proposed to provide multi-views [1, 3, 4] but they do no meet the previous requirements.

We propose to extend the concept of view point in a conceptual framework integrating modularity and abstraction to structure the process. Since "Software processes are software too" [2], it is possible to reuse technology and concepts borrowed from languages (modularity, encapsulation) and Software Engineering (configuration, abstraction), still retaining the point of view approach.

A software process is in fact the aggregation of numerous process fragments; each fragment describes different part of the overall process, with no overlapping. It is the usual approach, but it is far to be the reality.

In any large organization, each actor, or class of actors, has a *partial and overlapping* view of the complete process. This view corresponds to the process part this agent needs to be aware of, and should be expressed in the terms this agent is familiar with (its universe of discourse).

Most processes are multi-agent processes, i.e. *multiple agents* are collaborating in a single process, as it is the case in meetings, but also in most activities, like change request (involving both team leaders, designers, developers, validators and so on). It is a common oversimplification to consider activities as being independent and undertaken by a single agent.

A view can also be an *abstraction* of the "real" enacted process; it "sees" a limited part of the complete process, and represents a common reality under a specific presentation. Quality and monitoring are often abstract views.

The concept of view has been involved in different computer technologies (process, design, DB, AI, SI and so on) but under many different definitions. In our work, assuming a process model exists, (called the reference process), a view is defined as a process definition where parts of the reference process are missing (to be ignored in that view), parts are renamed (to fit the concepts known under the view), and parts are abstracted (reducing the level of detail to what is relevant in that view). "Part" here refers both to product and activity aspects.

In our work,

- Views can *share process instances*,
- A view is *enactable* either to monitor existing processes (observation process) or to interact with existing processes (actor process).
- A view can be defined a *posteriori* on an existing enacting process.

No multi-view formalism meet these requirements. In Tempo [1], views are applied in the product only, in [3] views are not enactable, in MVPL [4], they do not share process instances.

Since views can overlap, there is a risk that different views may define the same activity inconsistently. Since views are enactable, the same activity can be carried out twice. It is not easy to ensure consistency between the different views [3, 4]. The following requirements are particularly important:

- Process overlap should be detected.
- It should not be possible to duplicate activities,
- It should not be possible to define the same process in an inconsistent way.

An example of our approach is the following. Let there be a validation process (valid), and a quality assurance (QA) process. The former executes the technical validation activities; the later sees a part of these technical activities and also records, for measure and traceability purposes, some selected events, computes some values and displays the corresponding results. The real validation process is the union of these two processes. Each process sees only a sub-set of the activities and artifacts: libraries, the debugger and object codes are seen only by the validator; effort drawing, resource allocation reports and history records are visible only to the QA process).

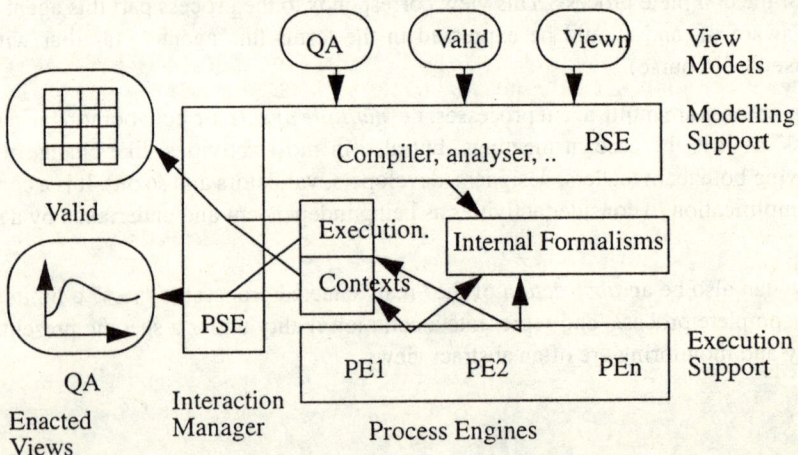

No formalism has currently achieved a consensus, and different Process Engines (PEs) are available, each of which focuses on a given aspect of software process modelling and support.

In the architecture we propose, views are modelled "independently" (see "The Méta-Process." on page 3) but all views are compiled together to produce the "real process model" in the Internal Formalisms. These internal formalisms are interpreted by different Process Engines which create and maintain different execution contexts for the internal enacting processes. For each defined view, the PSE is in charge of maintaining the corresponding monitoring and interaction interface, based on the view model and the different execution contexts.

2 Process Interface

Our approach is based on the fact that each process fragment has an interface. The concept of interface is borrowed from programming languages. A process interface is the visible part of a process; it defines the process functionalities both in an abstract non executable way and in a formal way; and the consumed and produced artifacts.

The private part contains the implementation. It describes the sub-processes needed, how sub-processes are composed (their ordering.) and coordinated, and how they cooperate. Collaboration (object synchronization) is supported by the Adele Work Space Manager [5], while cooperation is supported by Work Context [6].

3 The Méta-Process.

Our méta-process follows the following steps.

3.1 Model Building

A model can be built from existing process fragments (their interface). The goal here is essentially to reuse existing fragments.

The concepts of interface and process dependency allows technology developed for Software Configuration Management to be reused: automatic computation of process configuration which ensures the corresponding model will be complete and consistent and that reuse of existing process fragments is maximized.

3.2 Defining Views on a Process.

Once the model has been built, it is worth defining the view(s) needed on this (real) process. At this point in time, the model is indeed a complete model. The agent(s) which will use that process may not need to see the complete process, but only an *executable abstraction* of it i.e. an operational view.

A view is a new process interface built starting from the real process interface, and *hiding* part of the interface. Some parts are *renamed;* other *composed*, i.e. a connex sub part of the process can be abstracted as a single entity, under a single name.

The concept of view, as defined here, has strong links, at least from the product point of view, with the concept of view and virtual object type as defined in recent work on OODB, [7, 8, 9, 10].

3.3 Making Views Operational.

A view is operational if the agent can interact with it as if it were a real process. It means the view must be enactable, and that all aspects, such as monitoring, guidance and performance, must be possible from a view as well as from the complete process.

It means that a bidirectional correspondence must be permanently established between the view and the real process. This correspondence is not trivial, especially when composition of the real process has been performed, and is sometimes impossible (see [9], for a discussion of this topic). If the view is an observatory view, there is no restriction, (the correspondence goes only from the real process to the abstract view); otherwise (actor view) strong restrictions have to be imposed.

3.4 Sharing Process Instances.

In most work, each process fragment, when instantiated, produces a new independent process instance. We think this is an oversimplified view. In fact, most views share process *instances*, in the same way as they share object instances. Obvious examples are meetings, where agents (i.e. processes) share the same process instance (the meeting) or a monitoring view.

In practice it means the same execution context is shared by different processes. Each agent "sees" the same execution context, and just notices some action as being carried out "automatically" when they are performed by another agent of the same process. Sharing processes (and not only the artifacts) is a natural and simple way of implementing cooperative work. Otherwise, explicit cooperation protocols are needed to inform each process of the progresses made by the other, thus introducing unneeded complexity, and worse, the observed process needs to be modified.

We must explicitly define whether activities are shared and private, in almost the same way as we have to define whether the information is shared or private data.

3.5 The Méta-process.

Once the previous issues have been dealt with, the creation of a new process will be undertaken using the following steps:

1. Look for the needed processes, querying/browsing the existing interfaces (3.1). If they exist:
   - Select the convenient view of the needed existing process(3.1).
   - If no views are convenient, build the needed view(3.2).
   - Make new views operational (3.3).
   - Define the sharing of process instances (3.4).
2. If new process fragments must be defined:
   - Define the new process fragments.
   - Define the convenient view of these fragments (3.2).

4 Current Status

In our Esprit project, PERFECT, we integrated Process Weaver (work flow and petri net based) [6] and Adele (data flow and event based) to create a Virtual Process Engine built form both PEs and a BMS protocol for the communication and coordination of the basic PEs. We then defined a high-level language (APEL which stands for Abstract Process Engine Language) which is compiled towards the corresponding internal formalisms of Adele (triggers) and Process Weaver (tokens and transitions). A part of the architecture presented in section 1 has been tested.

Previous work on the Tempo formalism focused on view-point, but from a product perspective.

We have already implemented, in our PERFECT Esprit project, the way to make different and independent PE collaborate, on a peer-to-peer basis, the way to define a higher-level language compiled toward more basic PEs, and the way to coordinate execution contexts. We have also resolved the view point issue, when limited to the product aspects.

We are now planning to experiments on how to manage views of processes and the sharing of process instances, as well as support for definition and for quality modelling, monitoring and process enhancement views.

5 Conclusion

This approach has the following features.

- Maximum **reuse** of existing models, as seen in step 1.
- Avoidance of **duplication** of process models. In step 2, the process creation involves the checking that no existing processes have a similar description.
- Detection of **inconsistent** description of the same processes. All views are consistent by construction, they only abstract some aspects, they cannot provide inconsistent descriptions.
- Process **sharing** is an explicit feature. It avoids to defining independent activities, with explicit and complex collaboration patterns, when in reality the same process instance is shared.
- **Addition** of or **evolution** of a view, both at model and instance level, does not impact on existing processes.
- **High level formalisms** can be used for view modelling, different lower level formalisms can be used transparently for enaction and execution support.

We believe that the current approaches are lacking with respect to all the current above features, and that this approach could represent significant progress. It is felt that the last three points in particular are fundamental, since experience shows they are the main ways in which processes evolve i.e. by refinement of existing processes (adding a view which adds finer grain sub-processes), and adding services around the core processes (adding control, monitoring, quality, and so on). This approach introduces great flexibility into the global process, since adding views can be done with minimal work (maximum reuse), and minimal impact on other processes (view independence). Much work remains to be done, however.

References

[1]   N. Belkhatir, J. Estublier, and W. Melo. *ADELE-TEMPO: An Environment to Support Process Modelling and Enaction*, volume 3 of *Advanced Software Development*, chapter 8, pages 187–217. John Willey and Son inc, Research Study Press, Tauton Somerset, England, 1994.

[2]   L. J. Osterweil. "Software processes are software too." In *Proc. of the 9th Int'l Conf. on Software Engineering*, Monterey, CA, March 30-April 2 1987.

[3]   A. Finkelstein, D. Gabbay, A. Hunter, J. Kramer, and B. Nuseibeh. "Inconsistency handling in multi perspective specification." In *Proc. ESEC 93, LNCS 717*, pages 84–99, Garmish, Germany, September 1993.

[4]   M. Verlage. "Multi-view modeling of software processes." In B. Warboy, editor, *Proc. of European Wokshop on Software Process Technology EWSPT3*, volume 635 of *LNCS*, pages 123–127, Villard de Lans, France, February 1994. Springer-Verlag.

[5]   J. Estublier. "The adele work space manager." Adele Technical Report, available bt ftp.imag.fr, July 1994.

[6]   C. Fernstrom. "Process Weaver: adding process support to Unix." In L. Osterweil, editor, *Proc. of the 2nd Int'l Conf. on the Software Process*, pages 12–26, Berlin, Germany, 25 – 26 February 1993. IEEE Computer Society Press.

[7]   E. Rudensteiner. "Multiview: A methodology for supporting multiple views in object oriented databases." In *Proc. of the 18th VLDB Conference*, Vancouver, Canada, 1992.

[8]   E. Bertino. "A view mechanism for object oriented databases." In *Proc. of Int. Conf. on Extending Database Technology*, Vienna, Austria, March 1992.

[9]   A. Geppert, A. Scherrer, and K. Dettrich. "Derived types and subschemas: Toward better support for logical data independence in object oriented data models." TR 93.27, Institut fur Informatik, Universitat Zurich, 199.

[10]   Q. Chen and M. Shan. "Abstract view object for multiple oodb integration." In *First JSSST Int. Symposium on Object Technology for Advanced System*, Kanarau, Japan, Nov 1993.

Decentralised Process Modelling

Bashar Nuseibeh   Jeff Kramer   Anthony Finkelstein   Ulf Leonhardt

Department of Computing
Imperial College
180 Queen's Gate
London, SW7 2BZ, UK
Email: {ban, jk, acwf, ul}@doc.ic.ac.uk

Abstract. In this paper, we advocate decentralised process modelling and suggest that understanding and modelling the development processes of individual development participants is the key to supporting collaborative development. Our approach relies on recognising individual developers' states ("situations") by analysing local development histories. Different situations can be used to trigger a variety of further development actions, such as consistency checks between process models of different development participants. We report on experience using regular expressions to specify particular situations and rules to associate actions with these situations.

1. Motivation and Background

A significant proportion of large software development projects involve the participation and collaboration of many development participants, who in turn may be physically distributed. Software process modelling and technology address a wide range of issues surrounding the specification and development of complex systems, including the description of the activities by which software is developed, and supporting this with automated tools [6].

While an understanding of the processes by which software systems are developed is valuable, we believe that understanding and describing "fine-grain" software processes is equally worthwhile, and an effective approach to tackling many of the problems associated with decentralised development is therefore particularly useful. By "fine-grain" we are referring to *(a) developer level* processes which describe the activities of individual development participants rather than organisations, and *(b) representation level* processes which can manipulate elements of representation schemes (e.g., specification languages) rather than treating them as "vanilla" objects - whose structure and content is irrelevant or unknown [11].

To perform fine-grain process modelling and reap its benefits, many issues of decentralised software development need to be addressed. These include the specification of coordination behaviour between both individual development participants and teams of developers. Ben-Shaul and Kaiser [1] for example, adopt an "international alliance" metaphor in which participating "countries" adhere to "treaties" (c.f. process models) and engage in "summits" (at which process models are enacted). This approach addresses the broad problem of decentralised development between teams of developers rather than individual participants. Engels and Groenewegen [3] propose an approach to specifying the coordinated behaviour of different "objects" using a formalism called Paradigm - which, in turn, uses a special kind of state transition diagrams. In our approach, described briefly below (and in more detail in [9]), we also use state transition diagrams to describe individual developer processes.

2. An Approach to Decentralised Process Modelling

Approach. Our approach is to represent individual process models locally by associating them with individual development participants. For each process, we maintain an ordered *work record* of development actions which also defines a particular sequence of process *states*. These states can be used to identify appropriate courses of actions. Typically, a specific course of actions will be appropriate not only for one state but for a set of similar states (which we call a *situation*)[1]. This "decision knowledge" can be expressed by *rules* which map situations to actions:

<situation><course of action>

where a <situation> is the pre-condition of the rule, and a <course of action> specifies what should be done once a decision has been made (we have omitted post-conditions for simplicity - these could be used in more elaborate applications, such as planning). In other work [4], we adopted a similar approach to inconsistency handling in multi-perspective specifications, in which we used rules of the form "Inconsistency implies Action" to specify how to behave in the presence of inconsistency.

Finally, enacting process models elicits one of three kinds of responses (courses of action): *informal guidance,* which may include help text, video clips, etc.; *specific recommendations,* which include a limited set of actions that a developer is advised to select/perform; and *automatic execution of specific actions,* which are only performed if a developer is aware of the consequences of these actions and is prepared to relinquish control over their execution. These three kinds of responses reflect the amount of knowledge developers have in any particular situation, and the degree of automation they wish to adopt/impose on a development process.

Implementation. We have implemented our approach by treating process models as finite state machines, represented in terms of regular expressions (thus allowing us to make use of a variety of powerful and efficient tools, which in turn facilitate the prototyping of tool support). Enacting our process models causes the work record associated with each process to be updated. These process models also analyse their respective work records in order to recognise (match) particular situations (states). Decision rules can then be used to trigger the appropriate response (action(s)).

Associated with our general approach is a communication protocol that supports asynchronous message passing. This is a necessary interaction infrastructure to support the decentralised process modelling we propose, and to facilitate distributed consistency management in this setting. The reader is referred to [9] for details.

Example. To demonstrate our approach, we introduce the notion of *tests*, which are acceptor automata defined by regular expressions. A sample test, T_A, can be expressed in the form:

T_A: .*D[^R]*$ not-successfully-checked-since-D

The regular expression in this test uses the basic constructs of regular expressions as used in Lex [13]. Thus, this test is matched if a D-event, but no subsequent R-event, can be found in the local (developer) work record. D and R denote actions or communication events that are part of a local process model.

A sample rule, R_1, maps a situation to a response:

---

1 The term *situation* is a variation of the term used in the NATURE project's process meta-model [8].

```
Rule:          R₁

Situation:     T_A ∧ ¬T_B ∧ ¬T_C

Response:      recommend: name-clash-check
```

where the situation in this case is a logical proposition combining a number of tests (T_A, T_B and T_C), which, in turn, is a pre-condition for the response - a recommendation to perform the consistency check `name-clash-check` (other responses include `display: <message>` or `do: <action>`).

3. Evaluation and Summary

Process modelling is the construction of abstract descriptions of the activities by which software is developed [5]. While global process models may be useful for describing global system development activities, we have suggested that a decentralised approach is more representative, and ultimately more useful, for modelling distributed and concurrent activities in collaborative software development projects. Individual developer process models are easier to construct, and better vehicles for the provision of development guidance.

Having reaped the benefits of decomposing complex global models into simpler local ones, we further proposed an approach for coordinating the activities of these different models, by managing consistency checks between these models. Such coordination is achieved by deriving development states from individual developer process models, and then using these to trigger the consistency checks, which proceed according to a well defined communication protocol.

We have implemented a simple prototype tool in Objectworks/Smalltalk™ which demonstrates our approach. The tool illustrates how states are derived from individual developer histories, and then demonstrates our protocol of communicating state machines to drive the global development process. We intend to integrate this prototype with *The Viewer* [10], an environment supporting the ViewPoints framework [7], which we have developed to support multi-perspective development. In particular, we would like to build on our work of expressing the relationships between multiple ViewPoints [12], in order to facilitate the process of coordination and collaboration in this setting [2].

Acknowledgements

We would like to thank Michael Goedicke for his feedback on our work. This work was partly funded by the Department of Trade and Industry as part of the ESF project, the UK EPSRC as part of the VOILA project, and the European Union as part of the ESPRIT Basic Research Action PROMOTER and the ISI project. A variety of papers describing our work are available by anonymous ftp from dse.doc.ic.ac.uk in directory dse-papers.

References

[1] Ben-Shaul, I. Z. and G. E. Kaiser (1994); "A Paradigm for Decentralised Process Modelling and its Realization in the Oz Environment"; *Proceedings of 16th International Conference on Software Engineering (ICSE-16)*, Sorrento, Italy, 16-21 May 1994, 179-188; IEEE Computer Society Press.

[2] Easterbrook, S., A. Finkelstein, J. Kramer and B. Nuseibeh (1994); "Coordinating Distributed ViewPoints: The Anatomy of a Consistency Check"; *Concurrent Engineering: Research and Applications*, 2(3): CERA Institute, USA.

[3] Engels, G. and L. P. J. Groenewegen (1992); "Specification of coordinated behaviour in the Software Development Process"; *Proceedings of Second European Workshop on Software Process Technology (EWSPT '92)*, Trondheim, Norway, September 1992, 58-60; LNCS, 635, Springer-Verlag.

[4] Finkelstein, A., D. Gabbay, A. Hunter, J. Kramer and B. Nuseibeh (1994); "Inconsistency Handling in Multi-Perspective Specifications"; *Transactions on Software Engineering*, 20(8): 569-578, August 1994; IEEE Computer Society Press.

[5] Finkelstein, A., J. Kramer and M. Hales (1992); "Process Modelling: A Critical Analysis"; *(In) Integrated Software Engineering with Reuse: Management and Techniques;* P. Walton and N. Maiden (Eds.); 137-148; Chapman & Hall and UNICOM, UK.

[6] Finkelstein, A., J. Kramer and B. Nuseibeh (Eds.) (1994); *Software Process Modelling and Technology*, Advanced Software Development Series, Research Studies Press Ltd. (Wiley), Somerset, UK.

[7] Finkelstein, A., J. Kramer, B. Nuseibeh, L. Finkelstein and M. Goedicke (1992); "Viewpoints: A Framework for Integrating Multiple Perspectives in System Development"; *International Journal of Software Engineering and Knowledge Engineering*, 2(1): 31-58, March 1992; World Scientific Publishing Co.

[8] Jarke, M., K. Pohl, C. Rolland and J. Schmitt (1994); "Experience-Based Method Evaluation and Improvement: A Process Modeling Approach"; *Project report,* 94-15; ESPRIT Nature 6353, RWTH, Aachen, Germany, September 1994.

[9] Leonhardt, U., A. Finkelstein, J. Kramer and B. Nuseibeh (1995); "Decentralised Process Enactment in a Multi-Perspective Development Environment"; *(to appear in) Proceedings of 17th International Conference of Software Engineering*, Seattle, Washington, USA, 14-18th April 1995, IEEE Computer Society Press.

[10] Nuseibeh, B. and A. Finkelstein (1992); "ViewPoints: A Vehicle for Method and Tool Integration"; *Proceedings of 5th International Workshop on Computer-Aided Software Engineering (CASE '92),* Montreal, Canada, 6-10th July 1992, 50-60; IEEE Computer Society Press.

[11] Nuseibeh, B., A. Finkelstein and J. Kramer (1993); "Fine-Grain Process Modelling"; *Proceedings of 7th International Workshop on Software Specification and Design (IWSSD-7)*, Redondo Beach, California, USA, 6-7 December 1993, 42-46; IEEE Computer Society Press.

[12] Nuseibeh, B., J. Kramer and A. Finkelstein (1994); "A Framework for Expressing the Relationships Between Multiple Views in Requirements Specification"; *Transactions on Software Engineering,* 20(10): 760-773, October 1994; IEEE Computer Society Press.

[13] _____ (1983); *UNIX Programmer's Manual*; 7th edition, Bell Laboratories, Inc., Murray Hill, New Jersey, USA.

Coordination by Behavioural Views and Communication Patterns

Luuk Groenewegen and Gregor Engels

Leiden University, Department of Computer Science
P.O. Box 9512, NL-2300 RA Leiden, The Netherlands
email: luuk,engels@wi.leidenuniv.nl

1   Introduction

In [2] a proposal for a new specification formalism was formulated, aiming at modelling software processes, including nonhuman as well as human actors. In two later papers, [3] and [4], the formalism, from then on called SOCCA, was successfully applied to some small part of the standard software process modelling case ISPW-6, see [7]. SOCCA is an example of a so-called ecclectic modelling approach, as it combines and consistently integrates different existing modelling formalisms. SOCCA uses EER-like object-oriented class diagrams for the static perspective, state transition diagrams for the dynamic perspective, and PARADIGM for the communication. SOCCA is still under development, for instance the object flow reflecting the process perspective is not yet mature, and also computer support for SOCCA still has to be developped.

During the development so far, SOCCA has been used and is being used for much larger problems than the small part of the ISPW-6 case mentioned above. For instance in [5] a very detailed SOCCA specification is presented for process transactions [11] as implemented in MERLIN [6]. Another example of a very detailed SOCCA specification is given in [12]. It presents the SOCCA model for the complete process change part of the ISPW-7 case [8]. To this aim a substantially larger part of the ISPW-6 case had to be covered than discussed in [4]. A third example of a large and complex SOCCA model can be found in [10] where SOCCA has been applied to a certain part of the maintenance phase of a concrete software process as it is in use within Philips.

From these and other large SOCCA specifications we learn that indeed these specifications are very accurate as expected. They enable one to observe the solution of the problem situation at the lowest level of detail, where all instances are separately represented. On the one hand this is a very useful property, as it renders a SOCCA model nearly executable. But on the other hand, a serious drawback of this richness in accuracy is that SOCCA models tend to get really large, very complex and very detailed. Especially the PARADIGM part of the SOCCA models is liable to large proportions, high complexity and rich details. This is necessary for the complete specification of the communication, as this communication itself is very complicated and as it has so many details. But it nevertheless seems very useful to be able to restrict the specification to a more essential description. In Section 2 we propose two concepts which might be helpful in attaining such a shorter, more essential description. And in Section 3 we present our ideas about the research necessary for formally defining these concepts and for successfully applying them.

2 Views on the behaviour and patterns in the cooperation

In data modelling one of the possibilities to produce better manageable models, is to use views, data views. Views do not so much reduce the complexity of the model. Like any divide-and-conquer approach, using a view is a temporary restriction of the problem situation to a subproblem, resulting in a correspondingly partial solution, whose integration in the whole solution is postponed until later.

Our idea with respect to the communication part of SOCCA is to introduce views on behaviour. By means of such a view a modeller, and also a user of the model, should be able to restrict her or his attention temporarily to a certain relevant part of the communication. In our opinion a view on behaviour essentially is generalization of that behaviour which is occasional for that part of the communication at a certain point in time. This opinion is based on the following observation.

The three earlier mentioned master theses [5], [10] and [12] have in common what we think is a very interesting phenomenon. Really different parts of the whole behaviour within the model display a certain communication. These communications are differing in their details, but the essential parts of them are the same. By introducing the view on these different parts of behaviour as an occasional but common generalization of the behaviour parts, the common essence of that communication is made more explicitly visible. This certainly will clarify the complete description of the communication.

These views on the behaviour are called *behavioural views*, and the point of view is always from a hypothetic coordinator responsible for a certain part of the communication at a certain time.

Once we are able to specify, by means of behavioural views, the common essence of different behaviours with respect to a certain communication at a certain point in time, we can start to study possible sequences of communication in the course of time. Such a sequence of communication will be expressed in terms of these behavioural views. It can very well be that in totally different parts of the complete behaviour one observes analogous sequences, possibly embedded within even more detailed sequences, of these behavioural views. With respect to the length of these sequences, we think it appropriate to discriminate between rather short ones, relevant for communication on a micro level, and rather long ones, relevant for communication on a macro level.

Examples of communication on the *micro level* are synchronous communication, asynchronous communication, the rendez-vous, the broadcast. These and similar concepts can be considered as *communication units*, the small, short-termed communication protocols expressed in terms of the behavioural views.

The communication on the *macro level* then can be expressed in terms of these communication units. We call such a long (sub)sequence of communication units a *communication pattern*. Examples of communication patterns are protocols for transaction management. These communication patterns have been also observed in [9] within the context of modelling the process of a meeting.

Being able to indicate common communication patterns within a SOCCA model will certainly further clarify a SOCCA specification. Moreover, similar to data structures like records and lists providing patterns for the data perspective, and similar to control structures like if-then-else and while-loop providing patterns for the (sequential) behaviour perspective, structuring concepts for the communication perspective complementing the behaviour perspective towards nonsequential, i.e. parallel behaviour, are then being provided by our behavioural views, communication units and communication patterns.

3 Necessary research

We aim at presenting a rigorous description of the notions of behavioural view and communication pattern. Only a thoroughly formal description of theses notions will enable us to verify whether a SOCCA model without these notions remains sufficiently the same when these notions are applied to that model.

Therefore a thoroughly formal description of SOCCA is needed. Separate parts of SOCCA already have a certain formal basis. However, as SOCCA is an ecclectic modelling approach still missing an integral formal description, this has to be done first. Analogous to the approach taken in [1] we are going to combine, by (partially) reformulating and integrating them, existing formal descriptions for the EER part of SOCCA, its STD part and its PARADIGM part. Then the behavioural views and the communication patterns will be defined on this formal basis.

Once the notions are formally clear, they will be applied to the various examples mentioned earlier, from which we initially got the intuitive idea to develop these notions, see [5], [9], [10] and [12].

Another interesting idea is to use the notion of communication pattern as a basis for one or more possible classifications of communication. For instance, instead of having either synchronous or asynchronous communication, we expect to be able to indicate various degrees of asynchronism. This probably offers new ways to relate different communication concepts as used in various programming or specification languages.

But also, comparable to ideas in [9], a stronger relation can be established with research from other fields, e.g. addressing communication in meetings and in conferences - both within and witout the context of groupware - or addressing communication in organizations and in business processes. Based on the above notions, distribution and coordination will be specified in a formal manner as well as in a structured manner.

References

1. J. Ebert, G. Engels: *Structural and Behavioural Views on OMT-Classes.* In E. Bertino, S. Urban (eds.): Proceedings International Symposium on Object-Oriented Methodologies and Systems (ISOOMS), Palermo, Italy, September 21-22, 1994, LNCS 858, Springer, Berlin 1994, 142-157
2. G. Engels, L.P.J. Groenewegen: *Specification of Coordinated Behaviour in the Software Development Process.* In J.C. Derniame (ed.): Proc. 2nd European Workshop on Software Process Technology (EWSPT92), LNCS 635, Springer, Berlin 1992, 58-60.
3. G. Engels, L.P.J. Groenewegen: *Specification of Coordinated Behaviour by SOCCA.* In B. Warboys (ed.): Proc. of the 3rd European Workshop on Software Process Technology (EWSPT '94), Grenoble (France), February 1994, Springer, LNCS 772, Berlin 1994, 128-151
4. G. Engels, L.P.J. Groenewegen: *SOCCA: Specifications of Coordinated and Co-operative Activities.* In A. Finkelstein, J. Kramer, B.A. Nuseibeh (eds.): Software Process Modelling and Technology, Research Studies Press, Taunton 1994, 71-102
5. J. Höppener: *The MERLIN Process Transactions Specified with SOCCA.* Master Thesis, Computer Sc. Dep., University of Leiden, 1994.
6. G. Junkermann, B. Peuschel, W. Schäfer, St. Wolf: *MERLIN: Supporting Cooperation in Software Development through a Knowledge-Based Environment.* In A. Finkelstein, J. Kramer, B.A. Nuseibeh (eds.): Software Process Modelling and Technology, Research Studies Press, Taunton 1994, 103-130.
7. M. Kellner, P. Feiler, A. Finkelstein, T. Katayama, L. Osterweil, M. Penedo, D. Rombach: *ISPW-6 Software Process Example.* In T. Katayama (ed.): Proc. 6th International Software Process Workshop: Support for the Software Process. IEEE Computer Society Press, 1991.
8. M. Kellner, P. Feiler, A. Finkelstein, T. Katayama, L. Osterweil, M. Penedo D. Rombach: *ISPW-7 Software Process Example.* In M. Kellner (ed.): Proc. 7th International Software Process Workshop, IEEE Computer Society Press, 1992.
9. J. Lonchamp: *A Collaborative Process-Centered Environment Kernel.* In G. Wijers, S. Brinkkemper, T. Wasserman (eds.): Advanced Information Systems Engineering (CAiSE '94), LNCS 811, Springer, Berlin 1994, 28-41.
10. M. Rijnbeek: The Software Process: Maintenance of the SPI, Master Thesis, Computer Sc. Dep., University of Leiden, 1995.
11. St. Wolf: Ein transaktionsbasierter Ansatz zur Unterstützung kooperativer Softwareentwicklung, Ph.D. Thesis, Computer Sc. Dep., University of Dortmund, 1994.
12. A. Wulms: Adaptive Software Process Modelling with SOCCA, Master Thesis, Computer Sc. Dep., University of Leiden, 1995.

Configuration of Situational Process Models: an Information Systems Engineering Perspective

Sjaak Brinkkemper, Frank Harmsen, Han Oei

Design Methodology Group, Centre for Telematics and Information Technology, University of Twente, P.O. Box 217, 7500 AE Enschede, Netherlands
E-mail: {sjbr I harmsen I oei} @cs.utwente.nl

Vision

Currently, one can observe a convergence of the basis technology for systems development environments.

– Client-server tool infrastructures due to high performance network support.
– Platforms with transparent operating systems, e.g. Unix based PCs and workstations.
– Standard graphical user-interfaces (X-windows, Motif) for fast, intuitive interaction.

These technical facilities offer opportunities for a new vision on systems development tools: a flexible CASE workstation. For each project specifically a CASE workstation should be configured with tool components for the complete systems development life cycle (i.e. requirements engineering and software engineering). The most suitable tools are to be selected according to the application type, the available development expertise and the target infrastructure. For example, an object-oriented project could be performed using a selection of techniques from OMT and Fusion, in conjunction with a GUI development kit and a C++ programming environment. For a knowledge based system one desires to use the KADS method together with a Prolog interpreter. The CASE workstation should support any kind of project based on so-called method engineering functionality, i.e. support for the selection, adaptation and composition of tool components, together with transformation and generation utilities.

This vision has inspired the Method Engineering research project of the Design Methodology Group at the University of Twente. In this short position paper we want to explain the main results of this project: the Computer Aided Method Engineering (CAME) tool and the Method Engineering Language: **MEL**. The main goals of this project are partially implied by the following position statements:

• Systems development should be supported by project-specific methods and tools, e.g. situational methods.
• Process models and product models of systems development are equally important for the configuration of situational methods.
• The perspectives of the various disciplines (e.g. information systems engineering, software process modelling, software engineering) should be seamlessly integrated to benefit mutually from each others' expertise.
• Controlled flexibility of systems development is achieved by flexible process guidance and adaptive product structures.

Critical to the support of engineering situational methods is the provision of standardised method building blocks, which are stored and retrievable from a so-called Method Base. Furthermore, a configuration process should be set up that guides the assembly of these building blocks into a situational method. The building blocks, called *method fragments*, are process models and products models being part of IS development methods. We distinguish *product fragments* and *process fragments*. Product fragments model the structures of the products of a systems development method: at a high abstraction level deliverables and diagrams; at a low abstraction level concepts and constraints. Process fragments are models of the development process. Process fragments can be either high level project strategies, called method outlines, or more detailed procedures to support the application of specification techniques.

CAME tool

Currently, we are developing Decamerone, a CAME tool that is based on and is used in conjunction with the meta-CASE tool Maestro II. The architecture of Decamerone is depicted in figure 1. The three functional components in the tool, the method administration tool, the method assembly tool and the generators provide complete support for a method configuration process. Output of this CAME tool is a situational CASE tool.

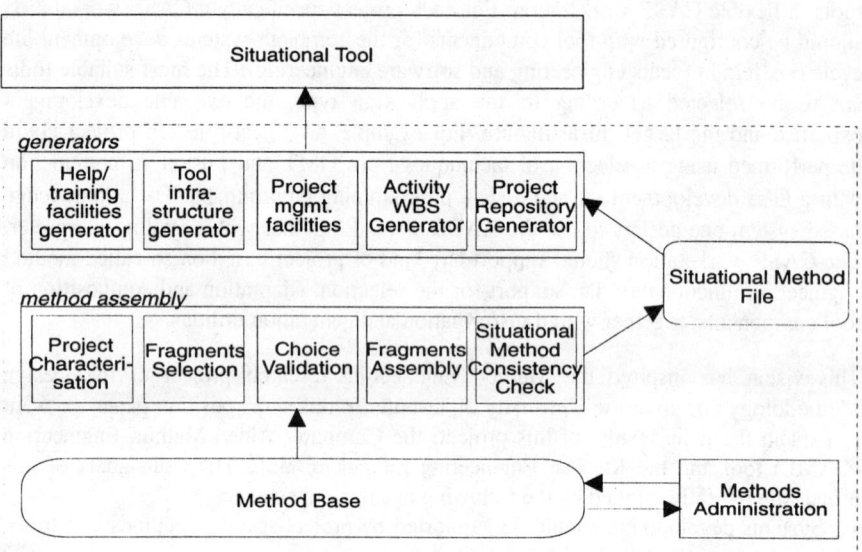

Figure 1   The architecture of Decamerone

Method Engineering Language

For description, administration, selection, and assembly of method fragments, we are developing the language **MEL**. **MEL** provides methodology-dedicated concepts and operators, which apply to both higher level method fragments, like stages and deliverables, and low level method fragments, such as concepts and their relationships. **MEL** descriptions can be represented graphically, but also textual or in a tabular form.

Method fragments are described by listing their components, and by specifying relationships with other method fragments. For process fragments, optionality, alternative steps, repeated steps, and parallelism can be specified. For product fragments, only optionality can be indicated. A large number of method fragment properties, such as *goal, purpose, creator, source method, application domain,* are keywords in **MEL**, possessing pre-determined value domains for ease of specification. To cope with method fragments derived from other method fragments (such as "logical data model" being derived from "data model"), and to enable multiple views on essentially the same method fragment (such as the manager's view on ERD and the analyst's view on ERD), an inheritance mechanism is introduced in **MEL**. Figure 2 and 3 show some examples of method fragments described in **MEL**.

```
PROCESS Perform Strategic analysis:
  LAYER Model;
  ID MRS/A3.1;
  SOURCE Merise;
  PARENT Obtain Strategic Plan;
  TYPE Abstraction;
  REQUIRED {Current Information system OPTIONAL};
  (
          - { Analyse organisation
            | Evaluate current flows};
          - Create Strategic analysis Report
  )
  DELIVERABLES {Strategic analysis Report}.
```

Figure 2   Description of a process fragment

Components of process fragments are either activity descriptions, decisions, or other process fragments, structured by constructs to model iteration, parallelism, optionality, and choices. In the example, the process fragment Perform Strategic Analysis consists of three sub-processes, which are elaborated in other process fragments. The required input and the produced deliverables of the process fragments are given by their names, and described in corresponding product fragments.

```
PRODUCT Current flow Diagram:
  LAYER Diagram;
  ID MRS/P4.2.3;
  SOURCE Merise;
  PART OF Strategic analysis Report;
  CREATED BY Evaluate current flows;
  (
          - Actor;
          - Flow
  ).

PRODUCT Actor:
  LAYER Concept;
  ID MRS/P4.2.3.1;
  SOURCE Merise;
  PART OF Current flow Diagram;
  ASSOCIATED AS is_source_of WITH Flow CARDINALITY (0,n);
```

ASSOCIATED AS is_target_of **WITH** Flow **CARDINALITY** (0,n).

```
PRODUCT Flow:
  LAYER Concept;
  ID MRS/P4.2.3.2;
  SOURCE Merise;
  PART OF Current flow Diagram;
  ASSOCIATED AS flows_from WITH Actor CARDINALITY (1,1);
  ASSOCIATED AS flows_to WITH Actor CARDINALITY (1,1);
  (
      RULE r1:
      flows_from(F₁,A₁) and flows_to(F₁, A₂) implies (A₁ ≠ A₂);
      # flows do not have the same activity as source and destination #
  ).
```

Figure 2b   Description of product fragments

The descriptions of products fragments contain, beside the usual properties, a list of all their constitutive diagrams and concepts. The concepts of the first product fragment Current Flow diagram are Actor and Flow, described by the other fragments. These concepts have a number of associations and syntactical rules.

Besides its ability to describe method fragment, **MEL** provides constructs for *method fragment selection*, by offering query operations, and for method *assembly*, by offering operations to combine or disconnect method fragments.

At present, the syntax and semantics of **MEL** are being formalised for the development of an interpreter for the language. Concurrently, we are experimenting with the configuration of method fragments into complete methods. This has resulted into a set of formal and semi-formal rules expressing the requirements on the consistency and the completeness of configured situational methods.

References

Harmsen F. and S. Brinkkemper, Description and Manipulation of Method Fragments for the Assembly of Situational methods. Memoranda Informatica 94-52, ISSN 0924-3755, 17 pages, October 1994. Submitted for publication.

Harmsen, F., S. Brinkkemper, H. Oei, Situational Method Engineering for Information System Project Approaches. In: A.A. Verrijn Stuart and T.W. Olle (Eds.), Methods and Associated Tools for the Information Systems Life Cycle. Proceedings of the IFIP WG 8.1 Working Conference, Maastricht, Netherlands, September 1994, IFIP Transactions A-55, North-Holland, 1994, ISBN 0-444-82074-4, pp. 169-194.

Harmsen, F., S. Brinkkemper, H. Oei, A Language and Tool for the Engineering of Situational Methods for Information Systems Development, In: J. Zupancic & S. Wrycza (Eds.), Proceedings of the ISD'94 Conference, Bled, Slovenia, September 1994.

Slooten, C. van, and S. Brinkkemper, A Method Engineering Approach to Information Systems Development. In the proceedings of the IFIP WG8.1 Working Conference on Information Systems Development Process. N. Prakash et al. (Eds.) Elsevier Science Publishers (A-30), pp. 167-186, September 1993.

Mechanisms for Cooperation (Chair: Christer Fernström)

Current Issues on Integration

Vincenzo Ambriola*   Giovanni A. Cignoni*   Christer Fernström**

*Dipartimento di Informatica, Università di Pisa, Italy   **Cap Gemini Innovation, France

Introduction

Much has been said about the need for powerful integration mechanisms in process-centred (software engineering) environments. In many cases, however, issues that have been raised stem more from a general wish for "total integrability" (where environment builders freely combine tools to work on any data produced by any tool) than from a careful analysis of what is really needed in terms of integration. In addition, many approaches build on (or try to cope with the deficiencies of) fairly low-level mechanisms, such as those available in standard UNIX systems. In order to achieve rapid progress in the deployment of software process technology, we suggest that researchers and practitioners more carefully analyse the minimal needs for integration and seriously consider how the emerging integration technology for personal computers can be applied in the context of process-centred environments.

Process-Centred Environments Needs

Software process technology helps supporting the software process by covering primary issues such as interaction with people involved in the process, the production and management of project documents, and the execution of development tools. In this perspective the ideal process enactment support is logically made of two components upon which the environment is built:

- the *process program interpreter* that supports the execution of the process activities as described with a process definition language;

- the *environment integration layer* that supports the management of the objects involved in the process (documents, products, development tools), their integration, and their integration with underlying networking systems.

Generally speaking, integration concerns interoperation between tools and documents. From the point of view of the software process, we are usually mainly interested in the aspects of coarse-grained integration (between tools and the process control mechanism), where data produced and updated at certain stages of the process is consistently made available to other parts of the process, while fine-grained integration (integration among tools) is much more an issue for tool set builders[1].

The design and the implementation of a language for process control is a well-defined problem. The knowledge of requirements and the available technology make this problem

---

[1] By tool set we mean a tightly integrated set of tools, typically used in combination, often by a single user, during an activity in the software process. Typical tool sets are design tool sets (design tool + design rule checkers) or programming tool sets (editor + compiler + debugger).

autonomous, guaranteeing a large degree of freedom. To date, many process modelling languages have been invented, proposed, realised, and even used.

The realisation of the environment integration layer, however, is generally a harder problem. In addition to the general concerns relevant to systems integration, such as concurrency control, networking, security, etc., the following issues (often underestimated) are constraints to the freedom of the environment implementation. In a more pessimistic view, they constitute serious impediments in the attempt of building an effective solution:

- The environment builder cannot choose the tools to be integrated, as they are imposed by the specific processes, and thus need to be introduced by the user organisation. Hence, the environment needs to be able to accommodate new tools and the integration cost must be small, in order to avoid that the set-up of the process environment be too heavy a burden on the process itself.

- A too loosely coupled integration can leave back-doors to escape the process; editors that allow the user to open documents independently from the process, tools that give access to other tools or to system shells.

- Interchange formats for documents are imposed by the adopted tools: this complicates the integration and the control of the flow of documents inside the process.

- The process is often based on heterogeneous platforms: for instance, in the same process the documentation could be maintained on personal computers while the development is performed on a network of workstations (with different UNIX flavours) or on a mainframe.

The majority of current research on process technology is aimed at producing demonstrators or experimental prototypes and, when analysed under the previous argumentation, show their weaknesses. Effective solutions for integration among tools, documents, process control and network resources in an environment suitable for process enactment do not seem to be available.

New Perspectives

We fully recognise that there are technical obstacles that slow the realisation of process-centred environments able to fully satisfy the integration requirements posed. However, the current impossibility to practically exploit the results of the research in software process technology is balanced by the existence (often ignored) of new, powerful, albeit less generic, mechanisms for integration that are the results of the continuous evolution of personal computing environments. A considerable de-facto standardisation has here taken place and the success of effective proposals for integration is encouraged by a wide community of users gradually educated to *electronic desk*: document processing (but without a process!), which is nowadays the most important activity performed on personal computer systems.

The idea of communication-based *software buses* started in the software engineering environments area and has been applied for service-oriented integration in different contexts (ESF, Field, SoftBench,...), and standardisation efforts have been undertaken to define standard services to be exported by different classes of software engineering tools (e.g. by Case Communiqué and Case Interoperability Alliance). Interestingly, the PC mass market has moved in a similar direction with the definition of DDE (Dynamic Data Exchange) by Microsoft and Apple Events by Apple: due to the fertile environment, these mechanisms are now widely accepted, exploited by application builders to offer new functions and available to end user programming via simple script languages.

Our view on the needs for integration from the perspective of process management is that of coarse-grained integration. From this perspective there is a need to provide different users with role-oriented pictures of information: a designer needs to provide design reviewers with pictures combining requirements, design rules and actual design elements; a software tester needs to be able to provide code correctors with combined pictures of functional specifications and results of test runs, etc. Rather than retrieving, analysing and reformatting the information needed to provide such combined pictures, users (and process enactment systems) need to be able to build "container documents" which accommodate hyper-text links to the information.

Due to the semantic and syntactical richness of the information manipulated by software engineering tools, it is however not feasible to burden the container document editors with the task of presenting and manipulating the information, but to rely on specific tools (e.g. those that originally produced the information) for this task. This approach to integration, sometimes referred to as "object linking" (the element referred to from a container document provides object characteristics by providing behaviour, such as "Display"), have in our meaning not been sufficiently explored in the context of process-centred environments. This is all the more surprising, given the recent trends on the personal computer market, where Microsoft's OLE (Object Link Embedding) and Apple's OpenDoc provide such mechanisms, which means that a large portion of software engineers already have access to the basic mechanisms on their personal workstations! Emerging versions of PC operating systems (like NT and Taligent) reinforce this kind of integration mechanism, while providing additional support to the systems integration issues of security, multi-user control, networking, etc.

Although they could provide part of the solution, it is clear that the functions provided by these mechanisms must be extended to cope with some of the specific needs that appear in (software) engineering. In addition to the general needs mentioned above, there are specific requirements that apply to object linking, e.g. the need to be able to handle versions and variants of data: the elements contained in (i.e. referred from) a container document usually need to be under version control and links will refer not only to an element, but to a version of an element.

In summary, our position is that tool and environment builders should explore the potential of the widely available integration mechanisms, while concentrating their efforts on providing additional support for the parts that are specific to (software) engineering.

Enveloping "Persistent" Tools for a Process-Centered Environment

Giuseppe Valetto[1] and Gail Kaiser[2]

[1] Rank Xerox Research Centre, "Le Quartz", 6, Chemin de Maupertuis, 38240 Meylan, France
[2] Columbia University, Department of Computer Science, 520 West 120th Street, New York, NY 10027, USA

1 Introduction

Process-centered environments usually support dialogues between external tools and the environment. There are three main categories:

White Box, where the tool's source code is directly modified to match the environment's interface. The changes can usually be implemented in a straightforward, repetitive manner, but nevertheless the source code must be available — a perhaps insurmountable difficulty when integrating COTS tools.

Grey Box, where the tool provides its own extension language or application programming interface in which functions can be written to interact with the environment. But relatively few tools provide such convenience. In principle, dynamic linking coupled with replacement of standard libraries might also work, if the framework's protocols are relatively simple.

Black Box, when only binary executables are available and there is no extension language. In this case, the environment must provide a protocol whereby *envelopes* extract objects/files from the internal representation in the environment's repository, present these objects/files to their "wrapped" tools in the appropriate format, and provide the reverse mapping for updated data and tool return values. Envelopes can also be used with Grey and White Box mechanisms.

The goal of our research is to extend enveloping principles to encompass a much wider array of tools than previously supported by the various systems that employ envelopes. We concentrate on the Black Box model, since it is often the only choice. Many process-centered environments and other kinds of environment frameworks have encapsulated tools following the Black Box model. The tools may or may not be connected to the environment via a message bus.

MARVEL's Shell Envelope Language (SEL) [3] is typical: A process administrator writes shell scripts augmented with special primitives that pass process and product data between the process server and the tool. Each envelope is associated with a process step. The process server sends the envelope's filename and its arguments to the user's client, which invokes the envelope. The script carries out any set-up operations, invokes the tool, and performs any cleanup after the tool terminates. It returns the results to the server, which executes the post-condition corresponding to the status code and continues the pending process fragment. This works well for UNIX utilities that accept all their arguments from the command line at invocation, read and write some files, and return a status code. However, there are numerous tools that don't fit this description:

Incremental tools, which request additional parameters and/or return (partial) results in the middle of their execution, such as multi-buffer text editors.

Interpretive tools, which maintain an in-memory state reflecting progress through a series of operations, e.g., Lisp systems. We are particularly concerned with permitting *different* users to submit activities to the *same* tool instance, even when that tool was not designed to support multiple distinct users.

Collaborative tools, which directly support interaction among multiple users.

We introduce a *Multi-Tool Protocol* (MTP) enveloping protocol, where *Multi* refers to submission of *multiple* activities to the same tool instance and enabling of *multiple* users to interact with the same tool instance. Tool instances may execute for an arbitrary period of time, far beyond the length of an individual activity by an individual user; thus we refer to the tool execution as *persistent*, analogous to how the UNIX operating system platform for SEL tools is persistent.

2 Tool Modeling

We modified MARVEL's tool declaration notation to include the new portion between square brackets ("[...]").

```
<tool-name> :: superclass TOOL;
  [ protocol  : (MTP, SEL) ;
    path : string ;
    architecture: (sun4, ...) ;
    host : string ;
    instances : integer ;
    multi-flag  : (UNI_QUEUE, MULTI_QUEUE,
                   UNI_NO_QUEUE, MULTI_NO_QUEUE)
    ; ]
  <activity-name> : string =
      "<envelope-name> <parameters locks>";
  <activity-name> : string =
      "<envelope-name> <parameters locks>";
  ... end
```

architecture gives the machine architecture on which the tool's envelope (and usually the tool itself) runs; alternatively, **host** gives the Internet address of a specific host when there is a restriction. **instances** specifies the maximum number of instances of the tool that can execute at the same time (0 means no upper limit).

multi-flag defines four categories of tools through two orthogonal dimensions: *UNI* vs. *MULTI*, where MULTI indicates that the same tool instance can be *shared* by several users, while UNI allows only for isolated work of each user with his/her own copy of the tool; and *QUEUE* vs. *NO_QUEUE*: where *simultaneous* execution of multiple activities (i.e., process steps) with respect to the same tool instance is supported for NO_QUEUE but not for QUEUE, whether UNI or MULTI. It may seem counterintuitive to think of these dimensions as orthogonal. In the case of MULTI_QUEUE, multiple steps on behalf of different users can share the same tool execution, but only one step actually runs at a time (FCFS); for UNI_NO_QUEUE, multiple steps can execute simultaneously in the same tool instance, but all must be on behalf of the same user.

path specifies the file system pathname of either the tool itself or of a shell script that runs the tool, while each **envelope-name** is another shell script that is invoked whenever the corresponding activity is executed. It handles the passing of arguments back and forth as well as interaction with a tool that is already running. For example, the envelope might indicate the text that the user should enter to the tool prompt or the menu item that should be clicked, with this information displayed in a special window to tell the user what to do. Such user intervention is necessary in the Black Box case, since we cannot assume any special facilities on the part of the tool for simulating user input, and redirecting "stdin" is insufficient for GUI tools.

MARVEL's process step syntax needed no changes. Different parameter sets to the same persistent tool instance result from separate instantiations of the same process step or from different steps. A persistent tool instance might receive requests from concurrent process fragments on behalf of the same or different users, or in sequence from the same fragment.

3 Sessions

We introduce tool *sessions*, beginning with an OPEN-TOOL command and ended by CLOSE-TOOL. A session's body is made up of a set of process steps determined dynamically as the process unfolds, where each activity maps to an individual process step. Although the activities are listed in sequence, they could potentially overlap, MTP-activityB beginning before MTP-activityA ends.

```
OPEN-TOOL <tool [session]>
    <MTP-activityA> <argumentsA> <session>
    <MTP-activityB> <argumentsB> <session>
    ...
CLOSE-TOOL <tool [session]>
```

The **tool** could refer to any tool declared as MTP. The optional **session** identifier distinguishes among simultaneously executing instances of the same persistent tool, so multiple users can participate in a session opened by another user (and, in principle, the same user could participate in multiple sessions for the same tool simultaneously, but this has not been implemented yet). Both arguments are selected from menus: when no instance has yet been activated, the OPEN_TOOL menu for that tool includes only the label *New*; newly created sessions are assigned identifiers that then appear in all user menus. It is possible to use an MTP tool without being compelled to issue the OPEN-TOOL and CLOSE-TOOL commands every time, as an implicit *atomic* session.

Imagine that USER1 opened SESSION1 for persistent tool TOOL1 and is executing MTP-activityA; then USER2 requests MTP-activityB, also on SESSION1.

- If TOOL1 is UNI_QUEUE or UNI_NO_QUEUE, USER2 cannot join the same session but can start a new session if the number of concurrent sessions for that tool would not exceed the number of instances allowed.
- If TOOL1 is not inherently multi-user (like most software development tools), but is declared MULTI_QUEUE, the system holds USER2's request in an *Activity Queue* until USER1 completes MTP-activityA. USER2 is not stuck waiting for this request to be processed, but may execute other process steps, or decide to abort and try again later. MARVEL already allows a user client to context-switch at will among in-progress process fragments.
- When TOOL1 is inherently multi-user, but not collaborative (e.g., a conventional database system), it would be declared as MULTI_NO_QUEUE; then, USER2's request is handled by the normal multi-tasking nature of TOOL1 — and USER1 and USER2 work in isolation. MARVEL's default concurrency control policy handles lock conflicts on data provided by the process server [1].
- If TOOL1 enables collaborative work, we again use the MULTI_NO_QUEUE attribute, but assume that most of the multi-user machinery is offered by the tool itself. Concurrency control on data provided by the process server must be relaxed to allow for the desired data sharing, e.g., two users might write the same file at the same time in a multi-user editor. MARVEL also supports process-specific concurrency control policies [4].

Note that sessions cannot be handled in the Black Box case by current broadcast message server (BMS) technology. A BMS approach allows for tools to signal events as they occur and for other tools to register for and receive notification of those events. But conventional Black Box enveloping would require all event receptions at the beginning of tool execution, typically to invoke the tool, and all event signalling at the end when the tool returns, analogous to the SEL protocol. However, when White Box or Grey Box tool integration is feasible, tools be modified or extended to support incremental reception/signalling of events, handling events from multiple users, etc. within a BMS framework.

4   Architecture

To implement MTP, we divided the process server's clients into new Special Purpose Clients (SPCs) and the General Purpose Clients (GPCs) inherited from MARVEL. GPCs continue to manage SEL envelopes.

SPCs do not need to interact directly with any human operator, so no user interface is needed; however, they need to manage the user input/output to/from persistent tools. This involves making the tools' own user interfaces available to the GPCs executing activities in the context of tool sessions. Most inherently multi-user tools are able to dispatch private instances of their interface to each user, but for other tools we exploited the public domain xmove utility [5], which transfers the GUI of a tool across workstations and X terminals. It is important to understand that simply resetting the X Windows `DISPLAY` variable would be insufficient, since the GUI instance has to start on one monitor for one user, then move to another monitor for a second user, etc. *without* reinitializing the tool.

SPCs also handle session bookkeeping. For example, when a `CLOSE-TOOL` command is invoked before the Activity Queue for that session becomes empty, the SPC automatically creates a new session and transfers the Activity Queue.

5 Conclusions

We have implemented MTP as part of MARVEL's successor, Oz, which adds a variety of other new functionality (see [2]). Example applications have included `idraw` as a UNI_QUEUE tool, where process steps are queued for one-at-a-time execution (the same userid may submit process steps from multiple clients, and the user interface is transferred as needed); `emacs` as a UNI_NO_QUEUE tool where steps are not queued but may overlap (typically on a single monitor); a local natural language processing system written in `commonlisp` as a MULTI_QUEUE tool, where steps are queued for one-at-at-time execution (and the UI is transferred among users participating in the same session as needed); and MARVEL itself as a MULTI_NO_QUEUE tool (that supplies its own clients for multiple users).

Mr. Valetto recently completed his MS thesis at Columbia University. Prof. Kaiser is supported in part by ARPA, NSF, AT&T Foundation, and Bull HN Information Systems. The views and conclusions contained in this document are those of the authors and should not be interpreted as representing the official policies, either expressed or implied, of ARPA, NSF, the US Government, AT&T, Bull or Xerox.

References

1. Barghouti, N.S.: Supporting cooperation in the MARVEL process-centered SDE. *5th ACM SIGSOFT Symposium on Software Development Environments* (December 1992) 21–31.
2. Ben-Shaul, I.Z., Kaiser, G.E.: A Paradigm for Decentralized Process Modeling and its Realization in the Oz Environment. *16th International Conference on Software Engineering* (May 1994) 179–188.
3. Gisi, M.A., Kaiser, G.E.: Extending a tool integration language. *1st International Conference on the Software Process: Manufacturing Complex Systems* (October 1991) 218–227.
4. Heineman, G.T.: Process modeling with cooperative agents. *3rd European Workshop on Software Process Technology* (February 1994) 75–89.
5. Solomita, E., Kempf, J., Duchamp, D.: Xmove: A pseudoserver for X window movement. *The X Resource* **1**(11) (July 1994) 143–170.

Coordination for process support is not enough!*

Yun Yang

CRC for Distributed Systems Technology
Faculty of Information Technology
Gardens Point Campus, Queensland University of Technology
GPO Box 2434, Brisbane, Australia 4001

1   Introduction

Any non-trivial software development needs to be carried out by teamwork, and as a result, software for group use has become more demanding. The increasing importance of computer-based teamwork support has been clarified recently in software development community [?, ?], and the corresponding software commonly developed are process centred environments. Currently, most process centred environments for software (as well as business) processes primarily focus on the **coordination** aspect. In reality, however, other functional aspects, i.e. **communication** and **collaboration**, are also critical to teamwork for process execution. This position paper asserts that research into process centred environments should concentrate on not only coordination, but also communication and collaboration.

2   Coordination in process centred environments

A process is commonly defined as a set of partially ordered steps intended to reach a goal [?]. Coordination supported by process centred environments focuses on sequencing these partially ordered steps in an automatic fashion. There are two perspectives seen for coordination which are similar to the concept "conversations *about* and *within* the process" termed in [?]:

- the activities that take place about the work, especially in a distributed environment
- the optimisation and support of the actual work performed

Process centred environments have been developed in both software and business process communities, namely, process-centred software development environments and workflow management systems respectively. At this stage we have seen around 30 process-centred software development environments prototyped

* The work reported in this paper has been funded in part by the Cooperative Research Centres Program through the Department of the Prime Minister and the Cabinet of Australia.

or released, and more than 140 companies claim that they are producing workflow management systems. A good comparison of the nature of these two processes can be found in [?].

In principle, a process centred environment is an automated system for interpreting the work of all software-related management and staff; it provides embedded support for an orderly and defined process [?]. We are fully aware of some ongoing research focuses on process modelling and evolution etc., however, discussions on and/or justifications of them are not the objective of this position paper. Although process research itself covers a very broad context, most prevailing process centred environments are very much concentrated on coordination functionality which can be viewed as a backbone for teamwork, as described in the next section.

3   Encapsulating communication and collaboration

In general, the cooperation involves three interrelated functional aspects, as advocated in the discipline of computer supported cooperative work (CSCW) [?], i.e., communication, collaboration and coordination. The coordination aspect forms a framework of cooperation enabling teamwork from the management viewpoint for sequencing steps of the whole process, but coordination itself only supports cooperation in a limited asynchronous manner. It does not provide **sufficient** and **effective** support for cooperation within each step when necessary. However, this can be significantly enhanced by encapsulation of communication and collaboration aspects for teamwork, as described in this section.

3.1   The communication aspect

Computer supported communication is normally based on the following two broadly divided scenarios to extend and enhance traditional communication:

- asynchronous communication
- synchronous communication

Asynchronous communication is normally supported by message systems which are useful for both informal and formal contact. For example, facilities for electronic mail are enabled and supported in many workflow management systems, and the formal message of "To-Do" lists can often be seen in process centred environments which has resulted from the existing coordination functionality.

Synchronous communication is supported by systems, for example, electronic tele-conferencing, which are useful for cooperation of activities such as co-designing (a program module) and reviewing (e.g., code inspection) in association with collaboration. So far, not much support of this kind has been enabled in process centred environments.

In summary, both asynchronous and synchronous communication facilities should be available in process centred environments. These communication facilities need to provide effective support for all team members.

3.2 The collaboration aspect

Effective collaboration demands that team members share information. Conventional database systems are useful, but **insulate** users from each other [?]. Finer granularity of information sharing from the following two perspectives is required:

- change flexibility which allows team members to update the shared information without much restriction
- change awareness which offers effective notification of each team member's actions when appropriate

An typical example is co-authoring, i.e. when a group of people are working cooperatively, it is necessary to allow them, sometimes simultaneously, to change shared information, such as updating a piece of code or writing review remarks. There are many mechanisms available for shared information updating such as the popular floor control policy, as employed in our review tool [?]. However, the ideal case would be no restriction which implies that any team member should be allowed to update any piece of shared information at any time, such as offered by the GROVE editor [?]. Consequently, some changes made must also be notified to others for consistency purposes, otherwise, cooperation would not be productive. Unfortunately, this issue has not been very well addressed in the literature.

Again, in summary, information sharing issues for collaboration should be handled satisfactorily by process centred environments.

4 Potential process support strategies

There are certainly different strategies for incorporating communication and collaboration aspects within coordination oriented process centred environments. For instance, on the support of asynchronous communication, we have seen some work on encapsulation of electronic mail facilities. For the experimentation and evaluation purpose, based on process centred environment Process Weaver [?], we have successfully integrated a locally developed Yarn tele-meeting system [?] which supports synchronous communication. This is useful for activities such as reviewing by a team. Moreover, integration of our review tool would be a helpful experimentation for further investigation of the collaboration aspect.

A more fundamental strategy may be to directly model all cooperation aspects into the process. However, as also implied in [?], this inevitably requires significant extensions to the existing process models. Therefore, along this way, there is still much to be done.

5 Conclusions

We have argued that the current process centred environments for software (as well as business) processes are essentially coordination oriented. Process support of coordination for cooperation by teamwork is not enough since communication and collaboration are also important and critical. We treat coordination support as a backbone for cooperation and envisage encapsulation of communication and collaboration support in process centred environments.

Acknowledgement

I am grateful for the help and advice of Jim Welsh, Geraldine Fitzpatrick, Glenn Smith, and Ken Baker.

References

1. B. Curtis, M. I. Kellner, and J. Over. Process modelling. *CACM*, 35(9):75–90, Sept. 1992.

2. C. A. Ellis, S. J. Gibbs, and G. L. Rein. Groupware: some issues and experiences. *CACM*, 34(1):39–58, Jan. 1991.

3. P. H. Feiler and W. S. Humphrey. Software development and enactment: concepts and definitions. In *Proc. 2nd Int. Conf. on Software Process*, pages 28–40, Berlin, Germany, Feb. 1993. IEEE Computer Society Press.

4. C. Fernström. Process WEAVER: adding process support to Unix. In *Proc. 2nd Int. Conf. on Software Process*, pages 12–26, Berlin, Germany, Feb. 1993. IEEE Computer Society Press.

5. V. Gruhn. Software process management and business process (re)engineering. In B. Warboys, editor, *Proc. 3rd European Workshop on Software Process Technology*, pages 250–253, Villard de Lance, France, Feb. 1994. Also in *Lecture Notes in Computer Science, Vol. 772*.

6. D. Heimbigner and M. Kellner. Software process example for ISPW7. Available by anonymous ftp from ftp.cs.colorado.edu, 1991.

7. J. Lonchamp. Supporting social interaction activities of software process. In J. C. Derniame, editor, *Proc. 2nd European Workshop on Software Process Technology*, pages 34–54, Trondheim, Norway, Sept. 1992. Also in *Lecture Notes in Computer Science, Vol. 635*.

8. M. Rees, G. Smith, R. Iannella, A. Lee, and T. Woo. Yarn: text based electronic meeting tools in a distributed environment. In *Proc. Computational support for distributed collaborative design – a workshop*, pages 3–20, Sydney, Australia, Sept. 1993.

9. W. Schäfer. Summary of the 7th Software Process Workshop. In *Proc. 7th International Software Process Workshop*, pages 28–31, Yountville, California, Oct. 1991. IEEE Computer Society Press, 1993.

10. J. Welsh, D. Spanevallo, and Y. Yang. A generic tool for document review by distributed teamwork. In G. K. Gupta, editor, *Proc. 17th Annual Computer Science Conference*, pages 679–686, Christchurch, New Zealand, Jan. 1994.

11. H. Wohlwend and S. Rosenbaum. Software improvements in an international company. In *Proc. 15th Int. Conf. on Software Engineering*, pages 212–220, Baltimore, Maryland, May 1993.

Coordination Theory and Software Process Technology

R. Mark Greenwood

Department of Computer Science
University of Manchester
Manchester M13 9PL, UK *
email: markg@cs.man.ac.uk

Abstract. Coordination theory is an interdisciplinary approach to studying the management of dependencies among activities. By its very nature software process technology deals with coordination. However it often expresses coordination in terms of low level details. An effective coordination theory would give us a better set of coordination abstractions. We illustrate the close relationship between these fields and propose areas where they could learn from each other.

1 Introduction

Coordination theory is the term used by Malone [9] to refer to the interdisciplinary study of coordination. It draws on a variety of different disciplines including computer science, organisation theory, management science, and economics. In these disciplines there are a variety of coordination problems and a range of solutions have been evolved to cope with them. One aim of coordination theory is to provide a language which can describe coordination problems and solutions in a way which allows them to be compared across disciplines.

Software Process Technology deals with coordination. By their very nature process centered software engineering environments coordinate their users and tools. Often, however, this coordination is achieved by low level details, the sending of individual messages. In general it is not possible to look at a process model and get an abstract view of its coordination dimension.

We will illustrate the close relationship between software process technology and coordination theory through an example. We hope that this will persuade the reader that the SPT community could both learn from, and contribute to, coordination theory. Finally, we make some tentative suggestions of benefits from understanding the relationship between these fields.

2 Coordination

Coordination is one of those ephemeral things which are best noticed by their absence. You spend a whole afternoon trying to find out who is supposed to

---

* previous address: Department of Electronics and Computer Science, University of Southampton, S017 1BJ, UK

be updating the xyz function with the extra parameter which you need, and eventually track down someone who says they thought that change was no longer required. This general feel is captured in Singh's definition: "Coordination is the integration and harmonious adjustment of individual work efforts towards the accomplishment of a larger goal" [10]. In [9] Malone proposes that:

Coordination is managing dependencies between activities.

and one way of progressing coordination theory would be to classify different type of dependencies, and identify the coordination processes used to deal with them. The following initial list of dependency types is given [9].

- Shared resources (including task assignments)
- Producer/consumer relationships
- Simultaneity constraints
- Task/subtask

Example coordination processes range from economic markets (a process to allocate scarce resources) to the mutual exclusion techniques of computer science.

A number of coordination languages have been proposed: Role Activity Diagrams [8], Diplans [7] and Role Interaction Nets [10]. These have been used to model organisational processes. In comparison with most languages applied to the software process, their focus is slightly different. Most software process modelling languages focus mainly on activities and their temporal relationships. The coordination languages tend to emphasise the allocation of activities into roles within an organisation, and the corresponding coordination requirements which arise from this allocation.

3 An Example

In [7] the "3 parts, 2 assemblies" problem is used to introduce Diplans. We will use it to give examples of the dependency types mentioned above.

The "3 parts, 2 assemblies" problem deals with the development of the following product family:

- Assembly aAC with component cA, component cC
- Assembly aBC with component cB, component cC

¿From this product structure is is easy to conceive of a generic development process:

1. three activities A,B and C to develop the three components, cA,cB and cC respectively.
2. two activities AC and BC to develop the assemblies with AC dependent on A and C, BC on B and C.

In this example there is an obvious producer/consumer dependency between the components and their respective assemblies. The simplest way of managing this is to decide that the assembly activities will wait for the completion of the respective component activities. We will formalise this coordination process using a simple process modelling language *coPML* [4, 5]. In *coPML* a process is modelled in terms of events, a state, and a set of constraints on the state. Our simple model will have 10 events, *startX* and *finishX* for each activity X, and the state will have 10 counts, #EV for each event *EV*. Our coordination process is captured by the constraints: there are 5 basic constraints, which reflect that activities start, finish, and are not repeated, and 4 constraints to cope with the producer/consumer dependencies:

$$0 \leq \#finishX \leq \#startX \leq 1 \; where \; X = A, B, C, AC, BC$$

$$\#startAC \leq \#finishA \quad and \quad \#startAC \leq \#finishC$$

$$\#startBC \leq \#finishB \quad and \quad \#startCB \leq \#finishC$$

A simultaneity dependency might mean tasks A and C cannot be done at the same time. We can amend our process to ensure this:

$$\#startA - \#finishA + \#startC - \#finishC \leq 1$$

A shared resource dependency might arise because there are only 2 people to do the development. If A and AC are allocated to person p1, B, C and BC to person p2, then B and C compete for p2's time. In this case we might want to prioritise C over B to reduce the chance of A being kept waiting:

$$\#startB \leq \#finishC$$

If we formulate the problem in terms of goals then it falls naturally into task/subtask dependencies rather than producer consumer ones. The assemblies aAC and aBC need to be developed, for aAC we need to develop cA and cC, for aBC we need cB and cC. We can imagine a different coPML model with constraints such as $achieveAC \Rightarrow achieveA \wedge achieveC$.

There are some types of coordination which are harder to express. In developing assembly aAC we might want a more concurrent approach with activities A and AC being worked on together. In this scenario AC could be kept up to date with the current state of cA, and able to request alterations which make developing aAC easier.

4   Possible Benefits

The above example illustrates that the software process technology community could contribute to the development of coordination theory. There are a large number, and variety, of dependencies between activities in the software process. Process modelling languages provide ways of expressing the coordination processes which manage these dependencies, and process enactment systems provide examples of exploiting the coordination potential of state-of-the-art technology. If SPT has much to contribute, what are the potential benefits?

There is nothing so practical as a good theory. [6]

We currently lack good coordination abstractions. This means that we consider coordination at a low level and prefer coordination processes on the basis that they are easier to implement rather than more appropriate. A good coordination theory would not only encourage process modellers to express their coordination requirements, it would also promote the recognition and reuse of process fragments. It might also help to relate coordination requirements to the networking facilities used to satisfy them.

Coordination theory could also help in the comparison of process modelling languages and enactment systems. The small example above illustrates that some dependencies arise from the allocation of activities to actors which is only weakly represented in coPML. In [7] it is emphasised that coordination involves both the temporal and this "spatial" dimension. The "spatial" dimension is the co-locating of the entities needed to do the work and the agent or actor. In PSS [1, 2] this would be dealt with by how activities are mapped to PML roles, in Process WEAVER [3] by the work contexts.

Another benefit would be a better understanding of the relationships between software process technology and other coordination disciplines. This includes other areas of computer science such as Computer Supported Cooperative Working (CSCW) and groupware. In addition, the interdisciplinary nature of coordination theory would help to clarify whether the type and number of dependencies between software process activities was significantly different from those for other processes.

Acknowledgements

Thanks are due to colleagues at the Informatics Process Group at Manchester University and the Declarative Systems Laboratory at Southampton University for many useful discussions.

References

1. R.F. Bruynooghe, R.M. Greenwood, I. Robertson, J. Sa, R.A. Snowdon, and B.C. Warboys. PADM: Towards a total process modelling system. In A. Finklestein, J. Kramer, and B. Nuseibeh, editors, *Software Process Modelling and Technology*, pages 293–334. Research Studies Press, 1994.
2. R.F. Bruynooghe, J.M. Parker, and J.S. Rowles. PSS: A system for process enactment. In *Proceedings of the First International Conference on the Software Process*, pages 142–158, Redondo Beach, California USA, October 1991.
3. C. Fernström. Process WEAVER: Adding process support to UNIX. In *Proc of the Second International Conference on the Software Process*, pages 12–26, Berlin, Germany, February 1993. IEEE Computer Society Press. sponsored by Rocky Mountain Institute of Software Engineering.
4. R.M. Greenwood. *Modelling Processes with Constraints*. PhD thesis, University of Southampton, 1994. in preparation.

5. R.M. Greenwood. Modelling processes with constraints. In B.C. Warboys, editor, *Software Process Technology: Third European Workshop, EWSPT'94*, pages 167–170, Villard de Lans, France, February 1994. Springer-Verlag. Lecture Notes in Computer Science 772.

6. C. Handy. *Understanding Organisations*. Penguin Books, fourth edition, 1993.

7. A.W. Holt. Diplans: A new language for the study and implementation of coordination. *ACM Transactions on Office Information Systems*, 6(2):109–125, April 1988.

8. A.W. Holt, H.R. Ramsey, and J.D. Grimes. Coordination system technology as the basis for a programming environment. *Electrical Communication*, 57(4):308–314, 1983.

9. T.W. Malone and K. Crowston. The interdisciplinary study of coordination. *ACM Computing Surveys*, 26(1):87–119, March 1994.

10. B. Singh. Interconnected roles (IR): A coordination model. Technical Report CT-084-92, MCC, 1992.

Transaction Technology for Process Modelling

Jens-Otto Larsen[1]* and Patricia Lago[2]

[1] Department of Computer Systems and Telematics,
Norwegian Institute of Technology, N-7034 Trondheim, Norway
[2] Dipartimento di Automatica e Informatica,
Politecnico di Torino, I-10129 Torino, Italy

1   Introduction

Process Modelling (PM) is a software engineering discipline that aims at supporting the development of large and complex software systems. Therefore, PM focusses on study, definition, enaction, and possibly improvement of software process models.

Traditional transaction management (TM) protects data against unexpected system failures, conflicts or inconsistencies. TM is also used in other areas, such as software engineering, to provide concurrency control and failure recovery for both automatic and manual activities. Transactions are typically used to provide reliable working contexts for software developers and for PM systems that enact process models. Software engineering TM adds requirements, such as support to user interaction, long lasting activities, non-programmed operations in the context of one transaction, and activity coordination.

In this paper we will look at some aspects of using transaction technology to support enaction of software process models. We define a simple transaction model and then discuss various ways transactions can be used in enacting the ISPW '6 PM example problem [KFF+90]. The focus of the discussion will be on structural aspects of both process models and transaction models. We identify problem areas, requirements to transaction models and choice-points in enacting a process model. By looking at process models, we find that more information about the enaction must be specified to be able to map it to a single execution of a set of transactions.

The structure of the paper is as follows: We first present some PM terminology and the ISPW '6 PM example problem. We then turn to transaction models and introduce a simple model typical for the class of transaction models used in PM systems. The next section discusses how parts of the example problem can be enacted in terms of issuing transactions against a database containing the software objects used in the example.

---

* Phone: +47 73594485, Fax: +47 73594466, E-mail: jensotto@idt.unit.no,
patricia@polito.it

2   PM and the ISPW '6 example

Most process model formalisms use the following concepts: *activity, product, role, person* and *tool*. An activity is used to model a logical unit of work, ranging from high-level tasks such as designing a system, to single tool invocations. Product items represent software artifacts as well as information about these. A role denotes the responsibilities and skills of the persons performing the tasks. Some PM formalisms have additional concepts, e.g. data-stores and control-flow.

A process model will then try to describe a development process in terms of these concepts and their inter-relationships, e.g.:

- Task – Product: A task has a number of input items, and outputs a set of new/changed items. The flow between tasks can be either direct or through a shared data store.
- Task – Person: A persons responsibility in carrying out a task is modeled by a role.
- Task – Task: Tasks may be related in several ways. There might be explicit constraints on the order in which tasks are enacted. In some formalisms, tasks may be refined into a number of steps each being a task on its own.

The ISPW '6 example problem describes a software change process. The tasks specified in the problem cover design, coding, testing and management. The product part of the example consists of various documents and source code and these are available either as documents or as files stored on a computer. The problem mentions a set of persons and their different responsibilities, e.g. the design engineer is both a designer and a design reviewer. In our discussion of the ISPW '6 example, we will focus on process steps (tasks), products and data flow between tasks.

We can classify the process steps in the problem according to several criteria. The process step `Develop Change and Test Unit` is an abstraction covering all the other process steps within the process, while `Modify Code` does not contain sub-steps. We will respectively call these types of steps *abstract* and *atomic*.

Referring to task duration, there are two kinds of tasks, (fig. 2): The first type lasts during the overall process, e.g. `Monitor Progress`, while the second starts when the input is available and stops when the output is produced, e.g. `Modify Design`. The first type of tasks are active in discrete intervals: The processing in the task is resumed upon notification reception and suspended when the output is produced.

Some process steps are not specified in full detail. The `Review Design` step involve several persons who will cooperate in some way to produce the output. The actual working process is left to the participants and it is not formally modeled.

We will consider two main types of product items described in the example: hand-carried documents, and item available on a computer. The problem specification gives a detailed explanation of the flow of hand-carried documents: these are taken from files and then either given to other steps or placed back in the

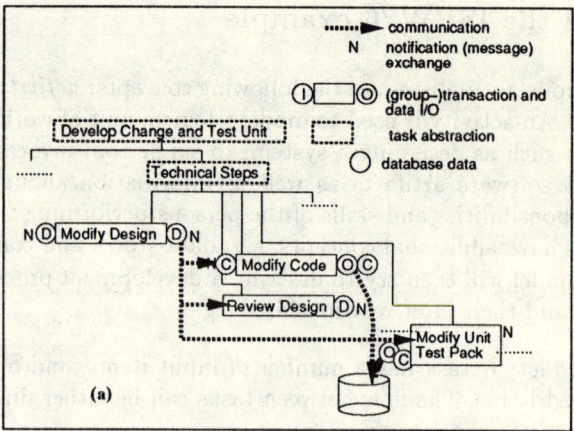

Fig. 1. ISPW '6: Process Communication/Structure

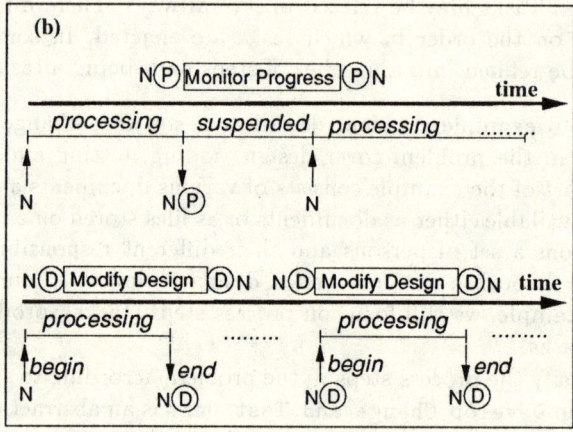

Fig. 2. ISPW '6: Process Duration

file. The example problem uses two types of document exchange between process steps (fig. 1): direct step–step (the modified design is carried from `Modify Design` to `Review Design`) and via shared storage (modified test plans from `Modify Test Plans` to `Modify Test Unit Package` via the test plans file). There is however a few open questions around the feedback cycle consisting of `Modify Design` and `Review Design`.

Turning to the product items which are stored on a computer, we do not find such a detailed explanation of flow and availability semantics, e.g. object code (fig. 1) and the modified unit test package. An example of an open issue is that the problem specification does not state whether data exchanged via shared storage is available outside the process before the whole `Develop Change and Test Unit` step (or one of its sub-steps) is finished.

3 Transaction Models

A transaction model can be characterised by transaction structure (transactions allowed by the model itself, from flat to hierarchical) and execution mechanisms (constraints on transaction behaviour).

Transaction models can also be described by their isolation properties, i.e. the ability to hide temporary results from other transactions during execution. Connected topics are:

Commit semantics: Some models make all changes available at the same time (commit), while in other models changed data can be made visible on an individual basis before the transaction finishes (pre-commit availability).

Local object space: In some models, objects are stored locally to the transaction until it commits, while in other models transactions work directly on the database, and are vulnerable to concurrent changes by other transactions.

Isolation is therefore dependent on how the above-mentioned characteristics are combined.

We will use a transaction model which is hierarchical, has standard commit semantics, and uses local object spaces. We do however need the possibility to communicate single objects to a designated set of sibling transactions, instead of the ordinary global commit.

4 PM Enaction and Transactions

In this section we describe some alternatives for mapping enaction of the different types of tasks defined in the ISPW '6 example to execution of transactions within the transaction model defined above.

We will assume that all documents used in the problem, not only the source and object code, are available on a computer and that they are stored in a database system supporting the transaction model we use. Synchronisation of the process steps will be done as part of the process, not necessarily using the locking features of the database.

We further assume that each atomic task described in the example problem is carried out within a transaction. When the persons responsible for a task are ready to start enaction, i.e. all input data are available and other constraints are met, they acquire the input data from the database, do the necessary processing and save the output data in the database.

When turning to abstract tasks, like process step `Develop Change and Test Unit`, we can choose between mapping this step to a transaction or not. In the first case, the other sub-steps are executed in sub-transactions, implying a hierarchical transaction model. In the latter case, the steps are enacted in transactions issued directly against the database. The problem does not state whether data exchanged via shared storage is available outside the process before the whole `Develop Change and Test Unit` step is finished. This means that both mapping approaches are valid, but in the second mapping alternative we cannot achieve isolation for the overall process. We can also note that transaction models using strict recovery will constrain the possible patterns of data exchange for both alternatives above.

When considering the duration of a process step, we see that there are two major ways to represent the first type of tasks (process-lasting) in a transaction model: We can either run one transaction for the whole life-time of the task, in which case the transaction model must allow transactions to make their changes available to other transactions before they terminate. The other approach is to start a new transaction whenever a task is resumed and commit the transaction when the output from the task is produced and task enaction is suspended. For the other type of tasks (shorter duration), we start a transaction when task enaction begins, and commit the transaction when it ends. If needed we can enclose feedback cycles in a transaction covering possibly repeated process steps.

When several persons are cooperating to accomplish a single step, e.g. `Review Design`, they should be able to able to use transactions to support their internally agreed-upon process (which will be ad-hoc in the global perspective). Two examples: The users can start a sub-transaction each and work in parallel, or they can agree on a serial access pattern for the common transaction corresponding to the task.

We have stated that the ISPW '6 example specifies several types of document exchange. To be able to support direct step–step exchange of data without making the data items available to other process steps, the transaction model must be extended with data-exchange operations. This applies both for flat and hierarchical transaction models.

Another issue related to transaction usage is the availability of data, which has implications on how various transaction features, e.g. locking, are used. In the problem specification there is no clear explanation of the availability of data neither during single steps, nor during the overall process. Some questions are: Are the input data of one task available to other tasks during its enaction? If so, to which tasks? (Alternatives are: No other tasks, to tasks within the same super-task, process, or globally.) Are all output objects available at the same time (when the step ends) or can data items be output before the process step ends?

5 Discussion

The ISPW '6 problem is described at a coarse-grained level. Existing software process modelling systems aim at automating parts of the process and use much more fine-grained process models.

In EPOS PM [COWL91] [JC93], tasks are used to represent all levels of activity, from projects to tool invocations by a single user. Tasks at the finest granularity are automatically planned and managed by the system, so that users do not need to be aware of all details.

Since all activities are represented by tasks in an EPOS process model and tasks are automatically enacted, it is not natural to enact all tasks in separate transactions. At the most fine-grained level, sets of tasks should be enacted in the same transaction, and there must be a way to define these groups of tasks. Since the process models are detailed, the problem of mapping task enaction to transactions is complex and should, if possible, be automated. The higher level tasks in EPOS can still be mapped onto transactions in the ways described in section 4.

By using a transaction model with structural properties similar to that of the PM formalism, the mapping becomes more natural. With weak isolation requirements, one may need to work around the rigidity of the transaction hierarchy to achieve sharing of data.

We have discussed problems and requirements arising when trying to map enaction of a coarse-grained process model to the execution of a set of transactions. We have identified some hooks that can be used for coupling enaction of process models to a transaction management system. There are also other ways to perform enaction in terms of executing transactions, and the choice of mapping is dependent on the desired level of isolation during enaction of both atomic and high-level tasks.

In general, the problem specification lacks a complete description of read, write, copy, and version semantics of input and output. This could be done in terms of availability of data or consumer/producer semantics. The questions posed above are related to the level of isolation that is required from the transaction model, and must be answered by PM systems that use transactions.

References

[COWL91] Reidar Conradi, Espen Osjord, Per H. Westby, and Chunnian Liu. Initial Software Process Management in EPOS. *Software Engineering Journal (Special Issue on Software process and its support)*, 6(5):275–284, September 1991.

[JC93] M. Letizia Jaccheri and Reidar Conradi. Techniques for Process Model Evolution in EPOS. *IEEE Trans. on Software Engineering*, pages 1145–1156, December 1993. (special issue on Process Model Evolution).

[KFF+90] Marc I. Kellner, Peter H. Feiler, Anthony Finkelstein, Takuya Katayama, Leon Osterweil, Maria Penedo, and H. Dieter Rombach. Software Process Modeling Problem (for ISPW6), August 1990.

Stepwise specification of interactive processes in COO

Claude Godart (1), D. Dietrich (2)

(1) CRIN-CNRS, ESSTIN, BP 239, F-54506 Vandoeuvre, France
(2) A.T.&T. Dataid, Immeuble BBC, 12, Rue D.F. Arago, F-57070 Metz, France
godart@loria.fr

Abstract. This paper deepens the idea of cooperation in the software process. It demonstrates the interest of distinguishing two levels of abstraction when modelling software processes to reason about cooperation in the sense of positive interactions between processes and respectively between the associated human agents. It illustrates the stepwise specification methodology used in the context of the the COO framework to specify process models.

1 Introduction

This paper deepens the idea of cooperation in the software process. It demonstrates the interest of distinguishing two levels of abstraction when modelling software processes to reason about cooperation in the sense of positive interactions between processes and respectively between the associated human agents. It illustrates the stepwise specification methodology used in the context of the the COO [1] framework and finally validates the practicality of the approach.

Section 2 illustrates our purpose through a motivating example. Section 3 specifies our software process model at two levels called respectively the conceptual level (3.1) and the logical (3.2) level, and describes the relationship between these two levels. This section must also be understood as a formalization of the idea of cooperation: this justifies its formal notations. Nevertheless, the next section (4) demonstrates that industrial people can access this formal work with medium effort and get benefits from this effort. Finally, section 5 concludes.

This paper formalizes the (an) idea of cooperation. For an intuitive introduction to COO see [Godart 94]; for a general overview see [Godart 93a] and shortly [Godart 93b]; about relationship between COO and the database field see [Canals 94a], about relationship between COO and CSCW, see [Canals 94b].

2 Motivations

Suppose the process *coding_task* has been defined as presented in figure 2.

This process has the general structure of a COO process. The root process represents the whole process which breaks down into sub-processes. At the leaves

---

[1] COO stands for COOperation ans COOrdination in the software process

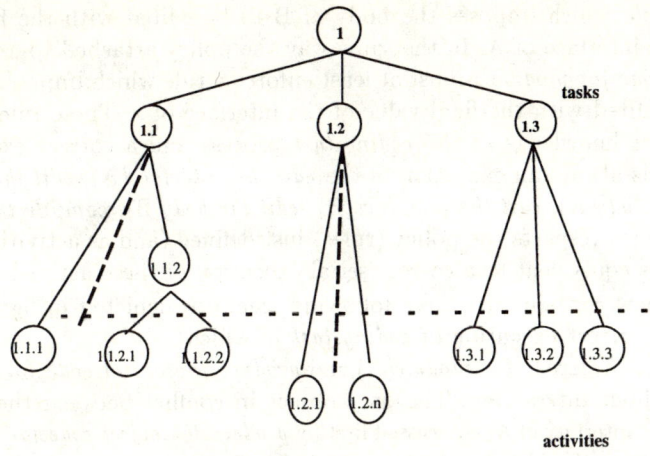

Fig. 1. Activities and tasks

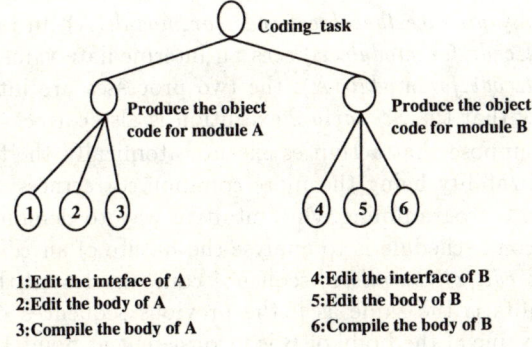

Fig. 2. A simple process description

are atomic processes called activities, at the upper levels are compound processes called tasks (see figure 1). When activities modify objects, tasks are only synchronization entities, they delegate object modifications to their enclosed activities.

This *coding_task* process has for objective to produce the object code for a module A and the object code for a module B. *coding_task* splits up into two tasks: *produce the object object for the module A* and *produce the object code for the module B*. Each of these tasks executes as a consistent combination of one or several occurrences of the *edit the interface, edit the body* and *compile* activity types applied to the interface or the body of A or B. Each activity is an abstraction for *read* and *write* operations. *edit the body*(B) is an abstraction for read the interface of A, read the interface of B, read the body of B and write the body of B. The fact that *edit the body*(B) reads the interface of A indicates that module B depends on module A. Thus, the the *coding_task* task must at least enforce a rule which imposes the body of B to be edited with the final (last) value of the interface of A. In the same way the policy attached to any *produce the object code for module x* must at least enforce a rule which imposes the body of x to be edited with the final value of the interface of x. These rules describe a part of our knowledge of the *coding_task* process and a correct execution of *coding_task* is always an execution of the *edit the interface*(A), *edit the body*(A), *compile the body*(A), *edit the interface*(B), *edit the body*(B), *compile the body*(B) activities which respects the policy (rules) just defined (and as activities execute atomically is equivalent to a correct serial execution of these activities).

Let us now consider the three following scenarios depicted in figure 3. *Scenario 1* is a correct execution of *coding_task* in which *produce_the_object_code_for_module*(A) and *produce_the_object_code_for_module*(B) execute without interaction. The only object in conflict between the two processes, i.e. the interface of A, is accessed first by *produce_the_object_code_for_module*(A), then by *produce_the_object_code_for_module*(B). Everythings occur as if *produce_the_object_code_for_module*(A) and *produce_the_object_code_for_module*(B) execute in sequence and in that order. At the opposite, in *Scenario 2*, *produce_the_object_code_for_module*(B) reads the interface of A in the same time it is being modified by *produce_the_object_code_for_module*(A). In fact, *produce_the_object_code_for_module*(B) sees an intermediate value of *produce_the_object_code_for_module*(A): the two processes are interactive. However, we can verify that the *Scenario 2* execution is also correct.

In fact, if we suppose that activities execute atomically, the first schedule is serializable, serializability being the more common correctness criteria[2] in the world of concurrent programming. One intuitive way to demonstrate the correctness of the second schedule is to analyse the nature of an *edit(x);compile(x)* sequence: the last *edit(x);compile(x)* sequence compensates all the previous (in the sense, its results is the same as if the previous sequences did not occur). Thus, even if the value of the body of B is inconsistent at point 1 because based on an intermediate value of the interface of A which may be inconsistent, if the final value of the interface of A is consistent, we can conclude that the final value of the body of B is also consistent (at point 2). And we have pointed out two executions of the same process which are both sequences of the same atomic activities and which are both correct. However, these schedules are not equivalent. The second allows interactions between *produce_the_object_code_for_module*(A) and *produce_the_object_code_for_module*(B) and must generally be favorized. In fact, interactions between activities also imply interactions between people and synergetic effects. These interactions are generally positive in social processes as software processes are.

It is clear also that in some organizations interactions can be considered as undesirable, and in some cases visibility of intermediate results must be prohibited, especially when operated by processes which cannot be easily compensated

---

[2] An execution of a process is serializable if it is equivalent (has the same effects on the objects) than a serial execution of its sub-processes. A serializable execution is correct.

```
         edit_the_interface(A);
           edit_the_body(A);
        compile_the_body(A);
         edit_the_interface(A);
         edit_the_interface(B);
           edit_the_body(A);
        compile_the_body(A);
         edit_the_interface(A);
           edit_the_body(A);
        compile_the_body(A);
           edit_the_body(B);
        compile_the_body(B);
           edit_the_body(B);
        compile_the_body(B);
```

Scenario 1: correct but without interaction execution of *coding_task*

```
         edit_the_interface(A);
         edit_the_interface(B);
       edit_the_body(B);    <-point 1
        compile_the_body(B);
           edit_the_body(A);
        compile_the_body(A);
         edit_the_interface(A);
           edit_the_body(A);
        compile_the_body(A);
         edit_the_interface(A);
           edit_the_body(B);
           edit_the_body(A);
        compile_the_body(A);
     compile_the_body(B);   <-point 2
```

Scenario 2: correct and with interactions execution of *coding_task*

```
         edit_the_interface(A);
         edit_the_interface(B);
           edit_the_body(B);
        compile_the_body(B);
           edit_the_body(A);
        compile_the_body(A);
         edit_the_interface(A);
           edit_the_body(A);
        compile_the_body(A);
           edit_the_body(B);
         edit_the_interface(A);
           edit_the_body(A);
        compile_the_body(A);
        compile_the_body(B);
```

Scenario 3: incorrect execution of *coding_task*

Fig. 3. Three different sequences of the same activities (ex example, putting a not enough tested product on the market can have huge consequences). In such a case, serializability of execution as in *Scenario 1* is the more commonly adopted solution. Finally, not that some sequences of the same atomic activities can be incorrect as *Scenario 3* demonstrates: the last value of the interface of A read by *edit_the_body*(B) is not the final.

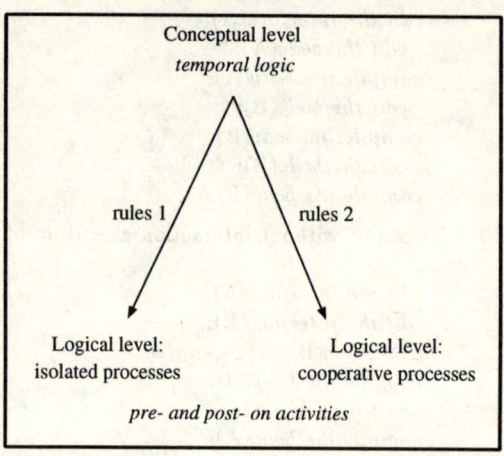

Fig. 4. Two levels of modelling

Our objective in this paper is to *characterize* non serializable but nevertheless correct executions (in fact in opposition with serializable executions but it was not our initial objective) and to propose a modelling technic which allows to specify processes in a way favorizing cooperation between human agents. The central idea in COO is that it exists an abstract level (called *conceptual level*, see 3.1) at which processes can be described independently of the nature of interactions. Then, incorporating interaction rules at the initiative of human agents defines different process descriptions (said to be of the *logical level*, see 3.2).

In fact, we have defined two sets of rules to transform one *conceptual* description either into a *logical*[3] description which allows serializable executions or into a *logical* description which allows cooperative execution (see figure 4). This specifies the (an) idea of cooperation in contrast with serializability.

3 The *COO* software process model

This section describes our software process model. Especially, we demonstrate how two interpretations (3.2) of the same constraint set can lead to cooperative

---

[3] Conceptual and logical can be understand with their common meaning in design methodologies: the conceptual level makes abstraction of any implementation consideration when the logical level results from some organizational choices.

executions (see 3.2) or serializable executions (see 3.2).

In addition, as the visibility of intermediate results is opportunistic to support cooperation when operations (operating intermediate results) are compensatable, visibility of intermediate results must be prohibited when these operations are not compensatable (it is difficult to reach again a consistent state when one of this operation has been launched in an inappropriate state). Assuming that serializability of executions is the more common way to prevent unacceptable visibility of intermediate results, in a realistic software process, if some parts can execute in a cooperative way, some must execute in a serializable way: we describe here how we are able to integrate "cooperative" process fragments with "serializable" process fragments.

3.1 Software process model (conceptual level)

As in [Thomas 89], we consider a software process model is a set of high levels goals and a description of how to decompose these goals into sub-goals. Thus, the execution state of a software process is the state of its goal tree and the set of data produced since its activation.

A software process as a 7-tuple In COO, a software process type is a 7-tuple $<S, V, H, P, O, I, C>$.

S : is a list of <parameter: type> couples called *Signature*. When the parameters of a process type are instantiated with actual values, a new process of the type is created,

V : is the *View*, the organization of data seen by processes of the type. This view is described by a set of objects types and relationships among these types in an Entity/Relationship manner,

H : represents the *Human agent* (one or several persons with the same or different role(s)) associated to each process of the type. The role of human agents is to choose which sub-processes must execute and when,

P : is the *Precondition*, a first order formula defining the workspace states in which a specific process can be started,

O : is the *Objective* (goal), defining the effects processes of the type have when they terminate (in fact, a set of "insert a relation", "delete a relation" and "update a function" clauses),

I : a set of process types called *implementation*. To execute a process p of type t, only processes of types in the *implementation* of t can be instantiated by the human agents associated to p. If its *implementation* is empty, the process type is called an *activity type*, if not it is called a *task* type. In addition, *init()* and *terminate()* activities are in the *implementation* set of all tasks. *init()* executes to start a task and to make initialization. the *init()* activity cannot execute if the precondition of the task is not validated. The *terminate()* activity executes inside a task to terminate this task: it cannot execute if this task has not reached its objective, has not the right effect on the object base, including the respect of constraints (see below). When termination of a task is rejected, it is the role of the human agent associated to the task to initiate new sub-processes in order to reach the objective of this task.

C : a set of constraints, called *integration constraints*. Integration constraints describe how sub-processes synchronize; in other terms, they are the basis for modelizing methodologies. We distinguish between *integrity constraints* and *transition constraints* (see 3.1 below).

The general idea is that the objective of a process describes its goal. The objectives of processes in its implementation are sub-goals of this goal. Thus, we obtain a tree of goals. Integration constraints and preconditions describe how sub-goals synchronize to contribute to the achievement of their father goal. As illustrated in 3.2, if goals are hierarchically organized, a process generally execute as a network of subgoals: two sibling processes can exchange intermediate results when executing.

More on constraints We use a limited form of temporal logic to specify constraints. We distinguish between *integrity constraints* which define *safety properties* between two states of an execution and *transition constraints* which defines *vivacity properties* between two such states.

Integrity constraints To express them, we use this special kind of temporal formula:

$$\forall x_1 : t_1, ..., x_k : t_k,$$
$$\Box(\alpha \supset \Box(\beta \; \mathcal{B} \; \gamma))$$

where

- x_1, x_2, ..., x_n are the variables of type t_1, t_2, ... t_n,
- $\Box$ means *"we have always"*,
- $\mathcal{B}$ means *before*,
- α is a non temporal and non disjonctive formula,
- β et γ are non temporal formula.

Intuitively, such a constraint is interpreted by "in all execution of a process of the type being defined, if an execution enters a state (of the object base) in which α is true, it is not possible for this execution to enter a state in which β is false until it enters a state in which γ is true".

Example:

$\forall$ y: bug,
$\Box((not\_fixed(\text{y}) \land$
$responsible\_for(\text{x}) = \text{z}) \supset$
$\Box(responsible\_for(\text{x}) = \text{z} \; \mathcal{B} \; fixed(\text{y})))$
forall x such that $bug\_on(\text{x,y})$

means that the people responsible for a source cannot change if it exists a bug report on this source.

Transition constraints To express transition constraints, we use this special kind of temporal formula:

$$\forall x_1 : t_1, ..., x_k : t_k,$$
$$\square(\alpha \supset \Diamond(\beta \; \mathcal{B} \; \gamma))$$

where x_1, ... , x_k, $\square$, α, β, γ have the same meaning than in the previous kind of formula and where $\Diamond$ means "*it is necessary*".

Intuitively, such a constraint is interpreted by "in all execution of a process of the type being defined, if an execution enters a state (of the object base) in which α is true, it is necessary for this execution to enter a state in which β is true before to enter a state in which γ is true".

Example:

$\forall$ i: interface,
$\square(new^4(i) \supset$
$(\Diamond \; new(b) \; \mathcal{B} \; terminated(@t)))$
forall b such that $body(used\_by(i,\_)) = b$

means that each time the interface of a module is modified, the body of any module which depends on this module need to be revisited before the current process terminates[5].

Semantics To define the semantics of constraints, we arrange the successive states of a process execution in an ascending order.

A state sequence σ satisfies an integrity constraint
$\Omega = \square(\alpha \supset \square(\beta \; \mathcal{B} \; \gamma))$, if, for all substitutions δ of the variables in Ω and for all state s_i of σ, i $\geq$ 0

if $s_{i,\delta} \models \alpha$ then
$\forall j, i \leq j < \mu,$
$s_{j,\delta} \models \beta$ with $\mu = min(\{k \mid k \geq i \wedge s_{k,\delta} \models \gamma\} \bigcup \infty)$

A state sequence σ satisfies a transition constraint
$\Omega = \square(\alpha \supset \Diamond(\beta \; \mathcal{B} \; \gamma))$, if, for all substitutions δ of the variables in Ω and for all state s_i of σ, i $\geq$ 0

---

[4] $new(x)$ is a function which return true each time a new value of x is produced in the workspace in which the constraint is evaluated

[5] in our framework, processes are represented as objects. Each object of type *process* has some attributes; especially, a unique identifier and a current state (invoked, active, suspended and terminated) as attribute. These objects can be used in α, β and γ formulas. @t is a variable which always contains the identifier of the task in the context of which the constraint is evaluated

if $s_{i,\delta} \models \alpha$ then
$\exists j, i \leq j < \mu,$
$s_{j,\delta} \models \beta$ with $\mu = min(\{k \mid k \geq i \wedge s_{k,\delta} \models \gamma\})$

Special case of constraints A special case of integrity or transition constraints exists when the formula α has the special value *init*. The semantic of such a constraint is the same than above, except that i has the value 0: *init* has the value true in the first state and only in the first state of the execution of the process which defines the constraint.

Justification If the other components of a description are rather classical in the domain, the way we express integration constraints is more unconventional. The decision to express constraints as logical expressions corresponds to the will to use an unified formalism to describe the differents aspects of processes. The choice of our special kind of temporal formula can be easily justified by the interactive and iterative nature of software processes: the properties we want to assert on the software products depend on the context in which we observe them, and especially on past and forecast states. Due to lack of space, we limit our justification to these two considerations, more items can be found in [Godart 93a].

3.2 The *Coo* software process model (logical level)

Constraints as just defined are not operational and need some transformation to be interpreted. Depending on the interpretation of integrity constraints, we will produce models which either favorize or prevent visibility of intermediate results. Before to describe how we implement constraints, it is requested to refine the idea of, on the one hand cooperative, on the other hand serializable, execution.

Correct cooperative execution of a software process

Closure of a task As in [Bancilhon 85], we introduce the idea of the *closure* of a task before to define a cooperative execution.

The closure of a task t $<s, v, h, p, o, i, c,>$ is a task t* $<s, v^*, h^*, p, o, i^*, c,>$ where:

- $i^*(t^*) = \{sp \mid (sp \in i(t) \wedge i(sp) = \emptyset) \vee \exists sp_i \in i(t) \wedge sp \in i^*(sp_i))\}$,
  $i^*(t^*)$ is the set of activities types (processes which do not break down) directly or indirectly enclosed in the task t.
- $v^*(t^*)$ integrates all the views of all the processes in $i^*(t^*)$[6],
- h^* is the union of all the agents playing the different roles in the task t.

---

[6] *view integration* is an open research topic, but not this of this paper: our framework is being developed atop the *Emeraude* system which implements the *PCTE 1.5* interfaces and *integration* has the meaning currently used in this world [Campbell 88]

Cooperative execution of a task A cooperative execution of a process t is a serializable execution of its directly or indirectly enclosed activities (of the activities in $i^*(t^*)$.

Note that this definition allows a task a to see the result of an activity enclosed in a task b before this task b terminates.

Correct cooperative execution of a task A correct cooperative execution of a task t is a cooperative execution of t which terminates (the objective of t is reached) in a consistent state (where integrity and transition constraints are satisfied) and in which each sub-process executes cooperatively and correctly.

Correct serializable executions of a task

Serializable execution of a task A serializable execution of a software process is an interleaved execution of its directly enclosed sub-processes which is equivalent (same results and same effects on the object base) to a serial execution of these sub-processes.

Correct serializable execution of a task A correct serializable execution of a task t is a serializable execution of t which terminates (the objective of t is reached) in a consistent state (where integrity and transition constraints are satisfied) and in which each execution of each sub-process is serializable and correct.

It can be surprising to suppose a serializable execution could be incorrect. That is due to the fact that in general schedulers make the hypothesis that each transaction implicitly respects the constraints or in other terms, constraints are implicit. Here, constraints are explicit and must be enforced.

Note also that a correct serializable execution is also a correct cooperative execution. This property allows us to integrate tasks which execute in a cooperative way with tasks which execute in a serializable way into the same process. In fact, a serializable execution is a cooperative execution with an additional constraint: the result of an activity which could not appear in a serializable execution cannot be shared between two active tasks.

Implementing constraints There is two important pending question in the two above definitions of correct executions: how to assure that " executions terminate in a consistent state (where integrity and transition constraints are satisfied) ?" and how to assure that "an execution is serializable from the point of view of integration constraints ?" (or in other terms, that a result which could not be seen in a serial execution, cannot be shared by two active tasks).

To answer these questions, we have developed two set of rules. The first set of rules transforms "conceptual" constraints into "logical" constraints which allow only correct cooperative executions, the second set into "logical" constraints which allow only serializable executions.

We content us here with a description of these transformations. However, they can be intuitively derived from our definitions of a correct cooperative execution and of a correct serializable execution (3.2). One important idea is also that a necessary condition for an execution to be serializable is that a process cannot start its execution in a state if starting its execution in this state risks violating a constraint. Validation of these transformations can be found in [Godart 93a].

Constraint transformations: principle We started from rules provided in [Lipeck 86] in the context of flat transactions. Correctness of our transformation can be found in [Godart 93a]. We make only one hypothesis: in case of serializable execution, the protocol ensuring serializability must be order preserving [Beeri 88].

Evaluating our constraints needs not only the knowledge of the current state of the object base, but also this of past states. To memorize these past states, we introduce for each type of constraints an additional predicate whose arguments are the variables appearing in the constraint description. This predicate must be entered as true in the database for a substitution ψ by the activity which validates a predicate α if:

- in case of an integrity constraint, the β formula must not be evaluated to true with that substitution in the next states until a state, in which γ is evaluated to true with that substitution, occurs,
- in case of a transition constraint, the β formula remains to be checked with that substitution in the next states before a state, in which γ is evaluated to true with that substitution, occurs.

We call such a predicate a *memo*. The basic idea we use to implement constraints is to extend the precondition and objective of activities to manage *memos* in a consistent way. The typical activities we consider are these which change at least one constraint component: α, β or γ.

Transformation of an integrity constraint: $\Box(\alpha \supset \Box(\beta B \gamma))$

In a first step, we introduce a predicate named, for the example, *Always Before* $(AB)^7$, whose arguments are the variables in α, β and γ. Then, if we suppose that the constraint is defined in the task t,

1. (a) in case of a cooperative execution, the objective of all the directly or indirectly enclosed activity types of t (in i*(t*)),
   (b) in case of a serializable execution
       i. the objective of all activity types in $i(t)$
       ii. the objective of the *terminate* activity of all task types in $i(t)$
   which can validate an α formula for a substitution is extended to give the value *true* to AB for this substitution (this is implemented by inserting a tuple representing this fact in the object base).
2. (a) in case of a cooperative execution,
       i. the objective of all activity types in $i(t)$ (the closure of t),

---

[7] it is clear that more application oriented names can be used in real processes

    ii. the objective of the *terminate* activity of all task types in $i(t)$

(b) in case of a serializable execution

    i. the precondition of all activity types in $i(t)$,

    ii. the precondition of the *init* activity of all task types in $i(t)$,

which can validate a β formula for a substitution, is extended to verify that the AB formula has not the value *true* for this substitution (to test that the corresponding tuple AB does not exist). The effect is to prevent an activity to violate a constraint,

3. (a) in case of a cooperative execution, the objective of all the directly or indirectly enclosed activity types of t (in $i^*(t^*)$),

   (b) in case of a serializable execution

    i. the objective of all activity types in $i(t)$

    ii. the objective of the *terminate* activity of all task types in $i(t)$

which can validate a γ formula for a substitution is extended to give the value *false* to AB for this substitution (this is implemented by deleting the corresponding tuple in the object base).

Example: transformation of the integrity constraint:

$\forall$ y: bug,

$\Box$((not_fixed(y) $\wedge$

responsible_for_modif(x) = z) $\supset$

$\Box$(responsible_for_modif(x) = z $\mathcal{B}$ fixed(y)))

forall x such that *bug_on*(x,y)

    (where x is a module and z is a person: this could be easily deduced, as the value of a z for a given x, from the object base. We do not describe this object base due to lack of place).

    In a first time, we introduce a predicate *responsible_cannot_be_modified*(x,z).

    Such a predicate is inserted (if not yet) by the process which produces a bug report on x, i.e, in case of a cooperative execution by an enclosed activity (of t^*), in case of a serializable execution, either by a directly enclosed activity (of t) or by a *terminate* activity of a task (if the process which produces the bug report is a task).

    If we suppose an activity *modify_responsible* (x:module, t:person). If *modify_responsible* is directly enclosed, its precondition is extended in all cases with the predicate

$\neg$*responsible_cannot_be_modified*(x,z)

If *modify_responsible* is not directly enclosed, in case of a cooperative execution, the *terminate* activity, of the sub-task of t which encloses it, in case of a serializable execution, the *init* activity of this sub-task, is extended with the same predicate.

    Finally, this predicate is deleted by the process which fixes the bug on x, i.e, in case of a cooperative execution by an enclosed activity (of t^*), in case of a serializable execution, either by a directly enclosed activity (of t) or by a *terminate* activity of a task (if the process which produces the bug report is a task).

Transformation of a transition constraint: $\Box(\alpha \supset \Diamond(\beta \mathcal{B} \gamma))$

The principle is the same as for integrity constraints except that the role of β and γ are exchanged. As example:

Transformation of a transition constraint Transformation of:
$\forall$ i: interface,
$\Box(new(i) \supset$
$(\Diamond\ new(b)\ \mathcal{B}\ terminated(@t)))$
forall b such that $body(used\_by(i, \_)) = b$

In a first time, we introduce a predicate *to_be_modified* (i, b).

Such a predicate is inserted (if not yet) by any process which modifies an interface and for each module which depends on this interface, i.e, in case of cooperative execution by any enclosed activity (of t^*), in case of a serializable execution, either by any directly enclosed activity (of t) or by the *terminate* activity of any task which modifies an interface.

Such a predicate is deleted by any process which modifies the body of a module and for all interfaces read to edit this module, i.e, in case of a cooperative execution by an enclosed activity (of t^*), in case of a serializable execution, either by a directly enclosed activity (of t) or by a *terminate* activity of a task (if the process which modifies the body of a module is a task).

Finally, the precondition of the *terminate* activity of the task t is extended with the predicate $\exists$ i, $\exists$ b | *to_be_modified*(i,b).

A conclusion Thus, the general idea is that in a serializable execution of a task, a sub-process cannot start its execution if starting this execution in the current state simply risks violating a constraint when the process will terminate. At the opposite, in a cooperative execution a sub-process can start its execution in any state, but it cannot terminate if its termination in the current state would violate a constraint. In other terms, in a serializable execution, consistency control is done *a priori*. In a cooperative execution, it is done *a posteriori*. See figure 5 as an illustration of this idea for the constraint "the Detailed Design cannot get a visa if the General Design has not" of the demonstration scenario implemented in *COO* (section 4.1).

4 Realism of the approach

Some people can question about the realism of our proposition. Apart the formalization effort, we have developed two actions to validate our work. First, we implicated a project manager from a software company in our group (in fact, the second author of the paper). Second, we have developed a framework on top of the *Emeraude* system to support cooperation during software development.

4.1 A realistic experience

Defining scenarios The second author of this paper is a project manager in a software engineering company who spent one year in our research team and

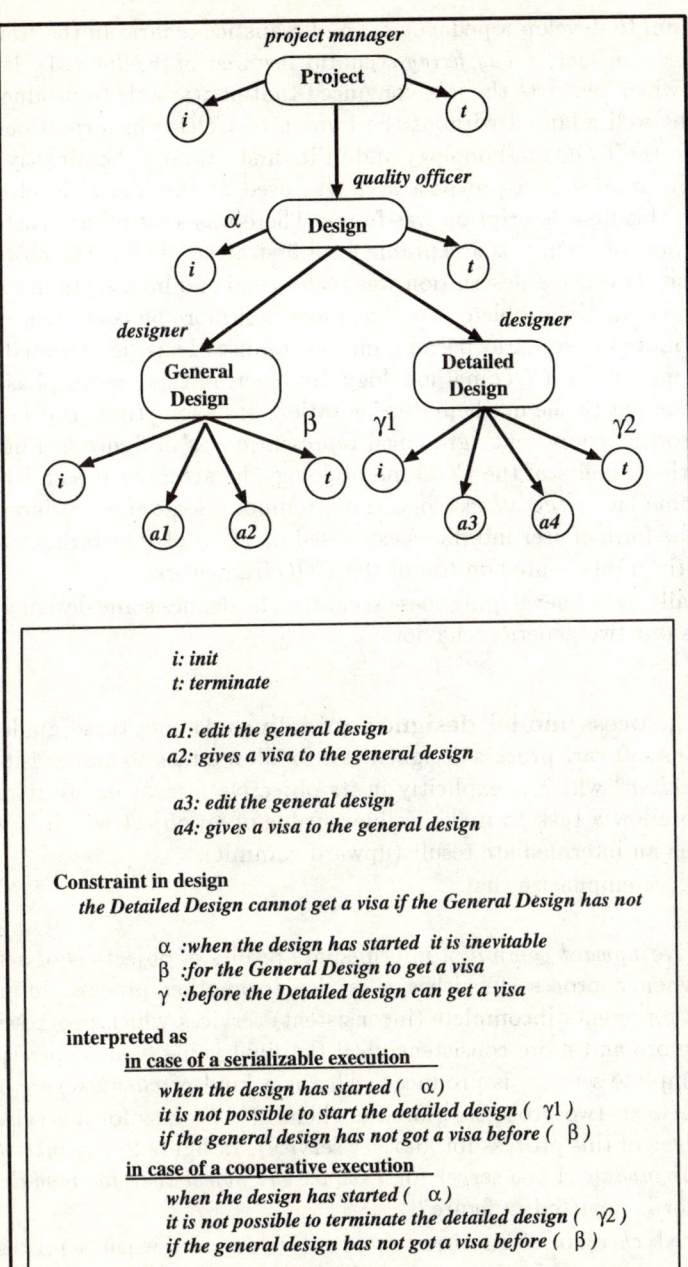

Fig. 5. Graphical description of constraints

we asked him to develop a pedagogical and realistic scenario in the frame of the *COO* project; in fact, a *bug fixing* scenario inspired of [Kellner 91]. It must be clear that when he starts the job, convinced that he was able to produce process descriptions well adapted without the burden of *COO* transformations, he was resistant to the *COO* methodology and in its first attempt, he directly specified its software process in a pre/post style (as used at the logical level of *COO*). Evaluating this first description was funny. There was a lot of inconsistency, but probably most of them was acceptable for a first attempt. But the more surprising was that its process description was really rigid and interventionist, allowing little interaction. His problem was that more and more he corrected errors and more and more his scenario became interventionist. Thus he accepted to make the experience of the *COO* methodology. However, in this second phase, he still remains resistant to the mathematical notations for a long time. But he invented him his proper "equivalent" graphical represention as in figure 5. Finally, convinced of the benefits of the COO methodology, he accepted it as a whole, may be with some last reservations concerning temporal logic. He produced its scenarios in the form of user interfaces expressed in TCL/Tk [Ousterhout 90] which are currently implemented on top of the *COO* framework.

Especially, when developing these scenarios, he defines some design guidelines and points out two generic behaviors.

Software process model designer guidelines Among these guidelines, he recommends software process designers not to allow a task to make visible a final result (*check_in*[8] which is explicitly in its objective (see 3) before it completes and not to allow a task to make visible a value of an object which is not in its objective as an intermediate result (upward commit).

In fact, we emphasize that

1. repetitive *upward commit* of intermediate results of objects is of active support when a process furnishes a service to another process: intermediate results represents incomplete (inconsistent) services which progressively become more and more consistent until the final value, which corresponds to the complete service, is produced. This is a kind of *client/server* organization between two processes (but the client of a process for a service can be the server of this process for another service). In figure 2, *Produce the object code for module A* is a server for *Produce the object code for module B*. This behavior is depicted in figure 6,

2. advanced *check_in* of an object is of active support when a process is not able to reach its objective without the intervention of another process. Typically, such an object is a bug report which is fixed by another process. We call this paradigm *client/mediator/server* paradigm in the sense that a mediator process is requested to allow a process to serve its clients requests.

---

[8] people not familiar with *check_in*, *check_out*, and *upward_commit* vocabulary can find some definitions in section 4.2.

In [Godart 93b], we give a scenario of *change control* (coming from [Kellner 91]) corresponding to this paradigm. This behavior is depited in figure 7.

4.2 The *COO* framework

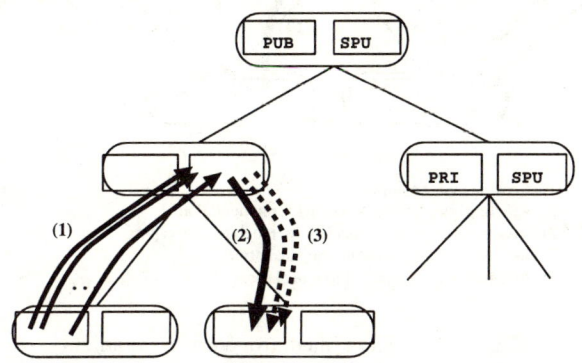

(1) **Repetitive upward_commit of an intermediate result**
(2) **First checking of an intermediate value of this result**
(3) **Refreshing of this intermediate result, at least with the final value**

Fig. 6. The client/server paradigm

We are implementing on top of the *P-RooT*[9] interface [Charoy 93, Charoy 94], an object orientation of the *Emeraude system* [Campbell 88], a protocol which enforces correctness of cooperative, partially serializable, executions (not which accepts all cooperative executions). For a current user, it appears as a workspace management system with transfer capabilities restricted to enforce consistency of executions.

Workspace management We distinguish between three levels of consistency for objects [Godart 93b]. *public* objects, when not being operated, respect all the constraints of all processes and can be concurrently accessed by all (sub) processes. *private* objects of a process, when not being operated, respect all the constraints imposed by this process and its sub-processes and can be concurrently accessed by all its sub-processes. *semi-public* objects of a process are objects which can momentarily violate some constraints imposed by the process and which can be read by (eventually by a restricted list of) sub-processes of this process. *public, semi-public* and *private* objects are respectively stored in *public, semi-public* and *private* object bases. There is only one public object base, but

---

[9] P-RooT stands for Pcte-Redesigned with object orientedTechnology

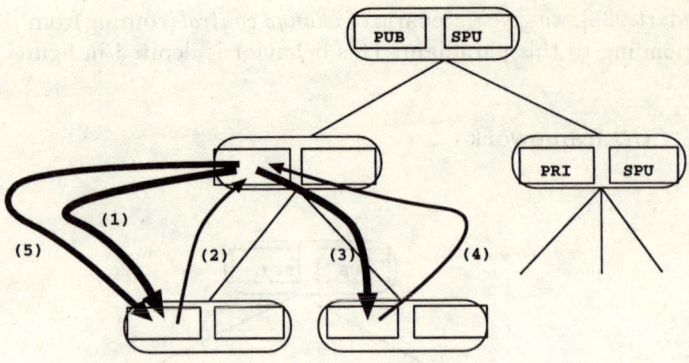

(1) checking_out_an_input
(2) reporting_on_this_input, asking for a mediation
(3) ckecking_out_what_is_nedded_to_react_to_the_report
(4) checking_in_the_result_of_the_mediation
(5) checking_out the result of the mediation

Fig. 7. The client/mediator/server paradigm

one private and one semi-public object base per task. Thus the *COO* development framework is based on a *base/sub-base architecture* [Kim 84], but flows of data between object bases are quite different, as demonstrated just below.

When a process want to read or write an object, it must transfer (may be simply logically, not physically), we say CHECK_OUT, this object into its proper private object base. This object is searched successively in:

- in its private object base,
- in the semi-public object base of its father process (if *read* mode),
- in the private object base of its father then in the public object base of the father of its father,
- and so on recursively in the object bases of its chain of ancestors until the public object base.

CHECK_IN is the reverse operation; a CHECK_IN activity transfers an object from the private object base of the process to the private object base of its father process. When a process completes, it CHECKs IN all its final results which have not been before. UPWARD COMMIT allows a process to transfer an intermediate result, i.e. a result of one of its directly or indirectly enclosed activity, from its private or semi-public object base to the semi-public object base of its father process. REFRESH allows a process which has read an intermediate result to refresh the value of this result with the new current value of the intermediate result.

Thus, semi-public objects of a task are results of activities which are not directly enclosed, or in other terms intermediate results of sub-tasks. In general, a sub-task continue to operate an object, an intermediate result of which has been produced.

Consistency maintenance rules

Concurrency control: basic rules Basically, a process executes as a nested transaction and, by default, the *nested 2 phases locking* mechanism assures safety of executions. In other terms, concurrency control is based on a locking mechanism. However, we accept exceptions to the *nested 2 phases locking* protocol. First, a task can make visible an intermediate result by UPWARD COMMITTING[10] it, but an intermediate result can only be read, and in addition, a process which has read an intermediate result of a task must read the corresponding final value (in case this intermediate result is modified again by the process which produced it). A task can also produce a final result before it completes, it can transfer it in the private object base of its direct ancestor. But in such a case, it cannot modify it again. We demonstrate in [Godart 93a] how these rules assure consistency of *memo*s insertion and deletion. These rules are additional rules to integration constraints checking as defined above, without forgetting the following remark.

Transfer activities and integration constraints Simply, note that transfer activities are activities as others and must be defined consistently with regards to constraints. Our position is that constraints are on final result, thus on CHECK_INed objects. In other terms, only CHECK_IN activities must be extended to insert and delete *memo*s. However, we have choose to extend also UPWARD_COMMIT activities to insert *memo*s; the fact that an intermediate result is always followed by a final result assures constraint maintenance. Note that these rules restrict the possible interaction with regard to transformation rules in 3.2. See [Godart 93a] for a large discussion of these choices.

5   Conclusion

This paper deepens the idea of software process model through one of its numerous facets: the interactions between processes and respectively the cooperation between the associated human agents. It is clear that it does not cover all the aspects of cooperation as defined in the *Computer Support Cooperative Work* paradigm and that we are concerned with long term gross grain processes.

It formalizes some aspects of the idea of cooperation and demonstrates that this formalization can be of real support when modelizing industrial software processes: distinguishing between a conceptual level and a logical level implies a stepwise specification methodology.

Work related to methodology to guide process development is not so numerous than work related to process enactment. Several of them exploit the

---

[10] in fact, upward commiting an object rest on the initiative of humanagents [Godart 93b]

properties of (extended) Petri-Nets [Bandinelli 94, Deiters 90]. Due to our model (rather rules based) our approach is more close to this of [Junkermann 94] and especially of [Sa 94] but with a focus on cooperation. We think also that management of intermediate results is of active support for process evolution [Conradi 92]. Our work is also related to (long term) transaction management. Entering in the details should enlight close relations with [Barghouti 92]. With respect to this work, this paper deepened one original aspect of the *COO* approach, i.e stepwise specification of transactions.

References

[Bancilhon 85]   F. Bancilhon, W. Kim and H. Korth. A Model for CAD Transactions. In *Proceedings of the 11th international conference on VLDB*, pages 25–33, Stockholm, august 1985.

[Bandinelli 94]   S. Bandinelli, M. Braga, A. Fuggetta and L. Lavazza. The architecture of spade-1 process centred see. In *EWSPT3*, Villard de Lans, 1994. LNCS772.

[Barghouti 92]   N. S. Barghouti. *Concurrency Control in Rule-Based Software Development Environments*. PhD thesis, Columbia University, 1992. Technical Report CUCS-001-92.

[Beeri 88]   C. Beeri, H-J. Scheck and G. Weikum. Multilevel transaction management: Theorical or practical need ? In LNCS 303, editor, *Advanced Database Technology Conference*, pages 134–154, march 1988.

[Campbell 88]   I. Campbell. Portable Common Tool Environment. *Computer Standard and Interfaces*, 8, 1988.

[Canals 94a]   G. Canals, F. Charoy, C. Godart and P. Molli. *P-Root & COO*: extending PCTE with new capabilities. In *ICSE Workshop on Databases for Software Engineering*, 1994.

[Canals 94b]   G. Canals, F. Charoy, C. Godart and P. Molli. The *COO* group. In *CSCW94 Workshop on Relationships between CSCW and Software Process research*, 1994.

[Charoy 93]   F. Charoy. An Object Oriented Layer on PCTE. In Ian Campbell, editor, *PCTE'93 conference*, November 1993.

[Charoy 94]   F. Charoy and P. Molli. Experimenting PCIS triggers on P-RooT. In *PCTE'94 conference*, November 1994. to appear.

[Conradi 92]   R. Conradi, L. Jaccheri, C. Mazzi, A. Aarsten and M. Nguyen. Design, use and implementation of SPELL, a language for software process modelling and evolution. In *EWSPT2*, Trondheim, 1992. LNCS635.

[Deiters 90]   W. Deiters and V. Gruhn. Managing Software Processes in Melmac. In *Proceedings of the 4th ACM SIGSOFT Symposium on Software Development Environments*, Irvine, California, USA, 1990.

[Godart 93a]   C. Godart. Contribution à la modélisation des procédés de fabrication de logiciel: support au travail coopératif (in french). In *thèse d'Etat (300 pages), Université de Nancy I*, 1993.

[Godart 93b]   C. Godart. COO: a Transaction Model to support COOperating software developers COOrdination. In *4th European Software Engineering Conference, Garmisch, LNCS 717*, 1993.

[Godart 94]   C. Godart, G. Canals, F. Charoy and P. Molli. An introduction to cooperative software development in COO. In *3nd International Conference on Systems Integration, IEEE Press*, 1994.

[Junkermann 94] G. Junkermann and W. Schaffer. A desigh metodology for process programming. In *EWSPT3*, Villard de Lans, 1994. LNCS772.

[Kellner 91]   M.I. Kellner and al. Ispw6 software process example. In M. Dowson, editor, *First International Conference On Software Process*, pages 176–186, California, 1991. IEEE Press.

[Kim 84]   W. Kim, R. Lorie, D. McNabb and W. Plouffe. A Transaction Mechanism for Engineering Design Databases. In *Proceedings of the 10th international conference on VLDB*, pages 355–362, Singapore, August 1984.

[Lipeck 86]   U.W. Lipeck. Stepwise specification of dynamic database behavior. *ACM SIGMOD*, pages 387–397, 1986.

[Ousterhout 90] J.K Ousterhout. Tcl: An Embeddable Command Language. In *Winter Usenix Conference Proceedings*, 1990.

[Sa 94]   J. Sa and B.C. Warboys. Modelling Processes Using a Stepwise Refinement Technique. In *EWSPT3*, Villard de Lans, 1994. LNCS772.

[Thomas 89]   Ian Thomas. The Software Process as a Goal-directed Activity. In *Fifth International Software Process Workshop*, 1989.

Session on Change and Meta-Process

Reidar Conradi, NTH

Software evolution (change) is pervasive, from initial development via maintenance to termination. The production processes driving such evolution are exposed to evolution too, since humans constitute their main operational agents.

It is common to break down a software process into a production process and a **meta-process**, that can change all parts of the process including itself.

The meta-process "operands" are both **external** process elements, such as production tools, human resources, company procedures etc., and **internal** (computer-supported) process model fragments describing the former.

Evolution can be "technical" according to life-cycle phases, e.g. to first define a general process model (or to make a project plan), customize it, execute this, and continuously revise it based on actual process performance. Evolution can also be "logical", capturing deeper process improvement wrt. productivity, lead-time, or quality. Such improvement assumes a *software experience database* to facilitate systematic project and organizational learning. Systematic measurements and reuse are important technologies here.

Process model evolution must be supported by proper methodologies (e.g. what are the decisive factors in making and evolving process models), formalisms (e.g. with reflectivity), and tools (e.g. flexible interpretation and dynamic binding) and related information databases.

The ideal process modelling language to support all the above is not made, and inter-operability and flexible tool architectures are increasingly important. We also need to regard the entire process model as a versioned, composite object, with conventional procedures for access control, traceability, impact analysis, and total quality.

The two papers in this session cover aspects of the above. The first is a full paper, the last a position paper:

- Jin Sa and Brian C. Warboys:
  "A Reflexive Formal Software Process Model".
  This reports work on a Software Method to support evolution through emphasis on a Reflexive Formal approach. An example is given.
- Ali B. Kaba, Jean-Claude Derniame:
  "Transient Change Processes in Process Centered Environments".
  This reports work on fine-grained process changes in such environments.

A Reflexive Formal Software Process Model

J. Sa
Department of Computing,
University of the West of England, Bristol
email: j-sa@csm.uwe.ac.uk

B.C. Warboys
Department of Computer Science,
The University of Manchester, Manchester
email: brian@cs.man.ac.uk

Abstract. In this paper a very simple reflexive formal software development method is described. The method is called OBM which provides a formal specification language. This paper demonstrates that the OBM development method can be defined as a process model in the OBM language. An example is used to illustrate how to develop and modify applications using the OBM development method.

1 Motivation for the Study

The study of software processes has attracted increasing attention in the last decade. One of the characteristics of such processes is that they are subject to frequent change. Thus, often either a software process needs to be changed or a new process needs to be created. The activities for managing a process, e.g. creating or changing, are usually referred to as meta-activities. A process consisting of the meta-activities for a particular process is referred to as a *meta-process*.

As pointed out in [6], since meta-processes are also processes, they themselves also need to be created and controlled. What is this meta-meta-process for producing a meta-process? Potentially, there is an infinite chain of meta-processes. This problem can be avoided if a process provides activities with which to manage itself. This kind of process is called a *reflexive* process. A model of such a process is called a reflexive model.

There have been several studies of reflexive process models. The aim of our study is to define a reflexive process model which has a formal mathematical foundation. Such a formal foundation offers the opportunity to be able to reason about the properties of a process model before the model is enacted.

2 Objective of the Paper

In this paper a very simple reflexive formal software development method is described. The method is called OBM which provides a formal specification language. In the rest of this paper, we sometimes refer to both the method and the language as OBM. The objective of this paper is to show that the OBM development method can be defined as a process model in the OBM language. Further more, this paper will illustrate how to use this model to develop an application model (which can either be the specification for a software product or a conventional process model), modify it or change the development process itself.

3 Background of OBM

Originally, OBM was developed to specify concurrent systems which may have interference behaviour. Among the available formalisms that were suitable for describing such systems, we decided to use temporal logic because it was capable of expressing liveness properties [13]. The particular temporal logic we chose offered a compositional approach [3]. Using that approach, a system is considered as a collection of components. Each component is specified as a temporal logic formula. The specification of the overall system is obtained by composing all the component specifications. We found that the specification notation was rich enough to express the properties of the types of concurrent system with which we were concerned. However the specifications were not very comprehensible. As a result, we gradually added more and more high level constructs to denote the temporal logic formulae. The collection of constructs was eventually evolved to a high level language with its semantics defined in the temporal logic.

According to the definitions given in [18], OBM may be classified as an object-based language. However our motivation for defining OBM was not to provide yet another object-oriented language, but to make it easier to specify systems using temporal logic.

In OBM, a process is considered to be composed of components which can be executed concurrently. The model of a process is developed at several levels. At each level, the model contains a collection of component specifications.

OBM provides *abstract objects* to model processes in a system at a level of abstraction, and *boxed objects* to compose a number of components and to hide the internal operations of this composition.

An *abstract object* models part of a system which may be refined later. It *provides* operations and may *require* operations provided by other components. The *interface* of an abstract object consists of a list of provided operations, required operations and an operation pattern defined in terms of the provided operations. Operation patterns are defined using an *ordering expression*. The detailed definition of ordering expressions can be found in [14]. The *body* of an abstract object consists of a *call template* for each required operation and a definition for each provided operation which contains its type, i.e. active or passive, its *call pattern*, and its *pre* and *post-conditions*. Each call template is of the form:

!*name*	
prop	*func/proc*
obj	*another*
apar	*v*

where !*name* is the name of the template, "prop" defines the type of the call, e.g. whether the caller receives a result, "obj" states the name of the component providing the required operation and "apar" specifies the actual parameter.

Call patterns are defined using ordering expressions, and specify the constraints on the ordering of the calls made to the required operations. Call patterns are defined in terms of call template names. Pre and post-conditions are defined in terms of the parameter, the result and the values received from the calls made to the required operations.

The behaviour of an abstract object is determined by the execution of its operations. The order of these executions must conform to the operation pattern. The temporal semantics of abstract objects are given in [11].

An abstract object may be *refined* by decomposing it into a number of sub-components. In order to justify the refinement, it is necessary to show that the composition of the sub-components is consistent with the abstract object.

The level i+1 refinement of a level i abstract object consists of:

1. *Component Refinement:* replacing the level i abstract object by a number of level i+1 components, and defining each level i+1 component.
2. *Operation Refinement:* replacing the sub-patterns of the level i operation pattern by operation patterns defined in terms of level i+1 operations. A sub-operation pattern may also be a single operation. Given a level i+1 refinement, a level i+1 operation is *internal* if it is not used in the *Operation Refinement* part.

The consistency checking of refinements is described in [11].

A *boxed object* is used for checking the consistency of refinements. It composes level i+1 components together to show that the composition is consistent with the level i abstract object. A boxed object contains a number of components and a list of internal operations. The behaviour of a boxed object is the concurrent executions of all the contained components. The behaviour of the internal operations is not visible. The temporal semantics of boxed objects are defined in [11].

4 The Reflexive OBM

Based on the original OBM, a new feature is added so that the OBM development process is reflexive.

Each object has two parts: the method part (or the meta-part) and the application part.

The method part defines how to develop the object. For example, the method part may describe that the first (meta-)operation is to specify the application part, after this, the application can be modified or refined, or the method part itself can be changed.

The application part is the result of applying the (meta-)operations in the method part. It defines the process to be modelled, for example, in this paper we use the example of a supermarket.

5 The Basic Reflexive OBM Development Process

Since the OBM development process is reflexive, the process itself can be changed. In this section, the basic (or the default) OBM process for developing an object (component) and its refinement is described.

To develop the application part of each component, the process is to:

- specify the component. Each component can be modelled as either an abstract object or a single object.
- verify that the specification is valid with respect to the specification in the level above.
- modify the specification of the component.
- refine the specification of the component. A refinement will create more OBM objects. Each object has two parts: the method part and the application part. The method part will be inherited from the parent when the object is created. Because this contains the ability to change self, the method can be modified and subsequently inherited by further refinements. The application part will be created using the method part.
- verify that a component's refinement is valid.
- change the OBM development process itself.

6 OBM definition of the OBM Development Process

Let's consider each OBM development process as an abstract object which provides several operations, for example, specify, modify, decompose and change self. The following is the definition of the interface of the application part. The operation pattern for the method is the same as the one for the application part.

Abstract Object $OBMdev$ The Interface Part:

provides

$$specify, verifyup, modify, decompose, verifydown, changeself$$

requires

$$none{:}none$$

operation pattern

$$specify; verifyup; (modify \oplus decompose; verifydown)^*; changeself; \circ$$

The above defines the interface part of the abstract object called *OBM dev*. It provides six operations: *specify*, *verifyup*, *modify*, *decompose*, *verifydown* and *changeself*. It does not require any operations. The allowed execution order is specified by an ordering expression.

The symbols used in the above ordering expression have the following meaning:

$P; Q$: P followed by Q.
$P \oplus Q$: either P or Q.
P^*: Zero or more occurrences of P.
∘: end.

The execution order of the above operation pattern is as follows: the first operation is *specify*; the specification must be verified against the level above; following the verification, there may be several iterations of one of two operations: *modify* and *decompose*; a decomposition must be verified; before each iteration, the object *OBM dev* may change itself.

7 An Example

In this example, we are going to explain how an application model can be developed and modified using OBM and how the basic OBM development process itself may be changed.

7.1 Specifying an Application Model using OBM

As mentioned in section 4, for each object in OBM, there are two parts: the method part and the application part. Let the operation pattern of the method part be the same as the one defined in section 6, i.e.

$$specify; verifyup; (modify \oplus decompose; verifydown)^*; changeself; \circ$$

Figure 1 illustrates that a system is to be developed. The enabled meta-operation is *specify*. (Enabled operations are underlined in the figures.) The application to be considered in this example is a supermarket system which deals with the back office management activities and the checkout activities. The supermarket system can be modelled as an abstract object which provides two operations: *management* and *trading*. These two operations can be repeatedly executed in parallel.

The supermarket model is developed or executed by executing the pattern of the method part. In this case, the first allowed (meta-)operation is *specify*. After doing *specify* we will have a definition of the application part. Let the definitionls be as follows:

```
specify; verifyup;
(modify +
decompose; verifydown)*;
changeself; ∘
```

Fig. 1. Starting Developing a System

Abstract Object *supermarket* <u>The Interface Part:</u>

provides

$$management, trading$$

requires

$$none:none$$

operation pattern

$$(management^* \| trading^*); \circ$$

7.2 Refining an Application Model

```
specify; verifyup;
(modify +
decompose; verifydown)*;
changeself; ∘
```

Supermarket

Fig. 2. After Specifying the Top Level

After specifying the top level, the allowed meta-operation is *verifyup*. Since this is the top level, the specification is by default valid. Figure 2 shows the situation after the application part has been specified and verified. To further develop the supermarket model, we continue to execute the method pattern. After *specify* and *verifyup*, the allowed meta-operations are: *Modify*, *decompose* or *changeself*. Assuming we choose to decompose the model.

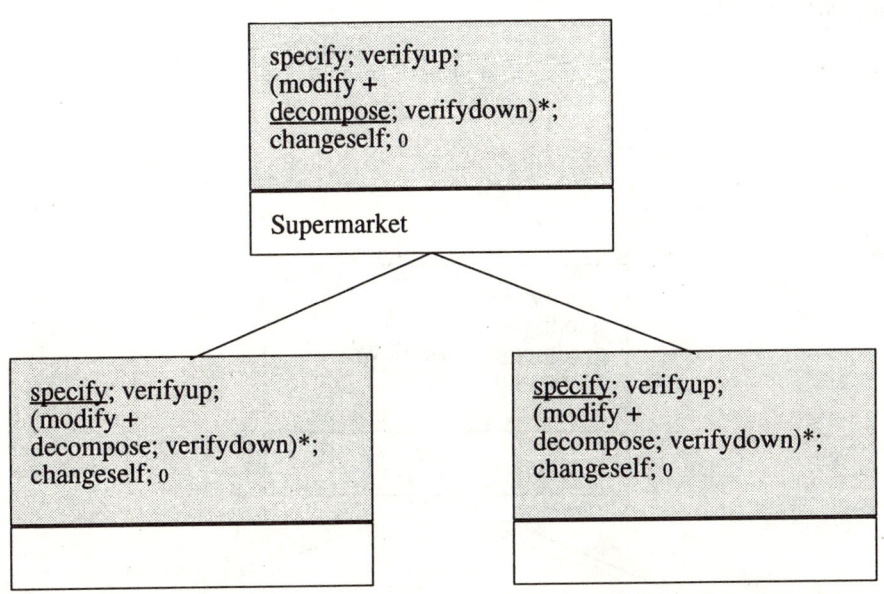

Fig. 3. Decomposing the Top Level

The operation *decompose* will create new abstract objects. In this particular case, the supermarket model is decomposed into two components: back office and checkout. So, two abstract objects are created. When a new object is created, the method part is inherited. See figure 3.

Consider the checkout component, it is defined as an abstract object. The pattern of the method part is

$$specify; verifyup; (modify \oplus decompose; verifydown)^*; changeself; \circ$$

Similar to the supermarket model, the checkout model is developed by executing the pattern of its method part. Again, the first operation is to specify the checkout model. The following is a specification of the checkout component.

Abstract Object *checkout* <u>The Interface Part:</u>

provides

$$login, servcust, logout$$

requires

$$backoffice{:}someoperations$$

operation pattern

$$(login; servcust^*; logout)^*; \circ$$

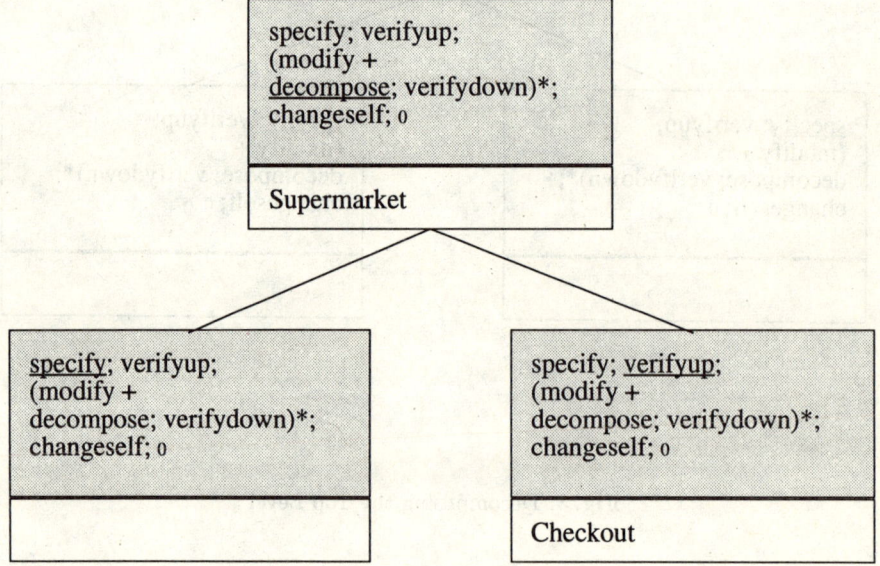

Fig. 4. After Specifying Checkout

The above specification defines that *checkout* provides three operations: *login*, *logout* and *servcust*. It requires some operations provided by the back office. The execution order specified by the operation pattern defines a repeated trading pattern. In each pattern, one is allowed to login, serve customers repeatedly, and then logout. See figure 4.

7.3 Verifying the Decomposition

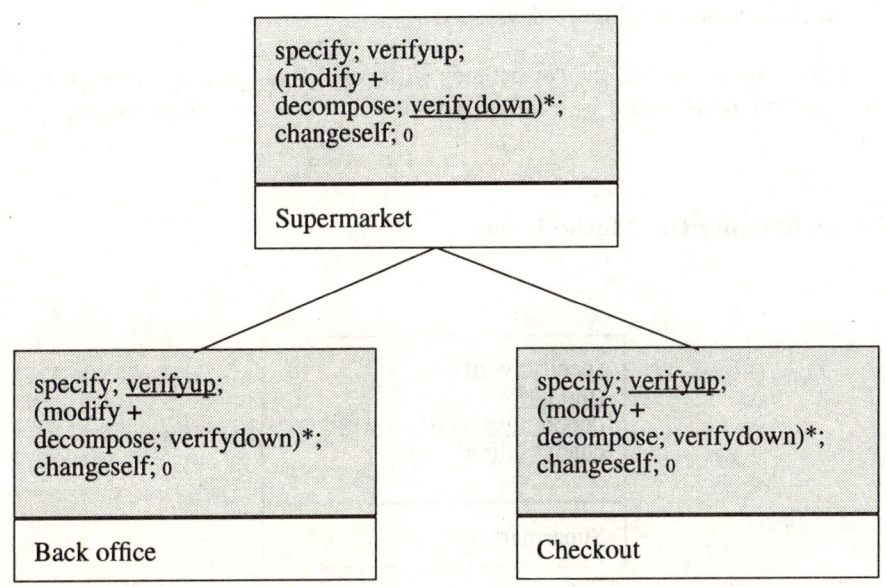

Fig. 5. Verifying the Decomposition

The decomposition must be checked for its consistency, i.e. we must prove that the composition of back office and check out is consistent with the supermarket node. This type of proof is described in [11]. Figure 5 shows that the supermarket node is performing the verification meta-operation. This operation calls the meta-operations *verifyup* of both back office and checkout.

7.4 Modifying an Application Model

Consider the checkout node, after specifying and verifying the checkout component, we can continue to develop the checkout model by executing the pattern of the method part. For example, we can use *modify* to change the operation pattern of the application part to

$$(login; (servcust; logtransaction)^*; logout)^*; \circ$$

With this new pattern, each service to a customer must be logged.

The modification of the application part of a node may trigger other changes:

− any node that uses this node, e.g. back office, may need to be changed
− all child nodes must be re-developed

After a modification, the consistency with respect to the level above, e.g. the supermarket node, must be checked again. The proof procedure is defined in [11].

7.5 Changing the Method Part

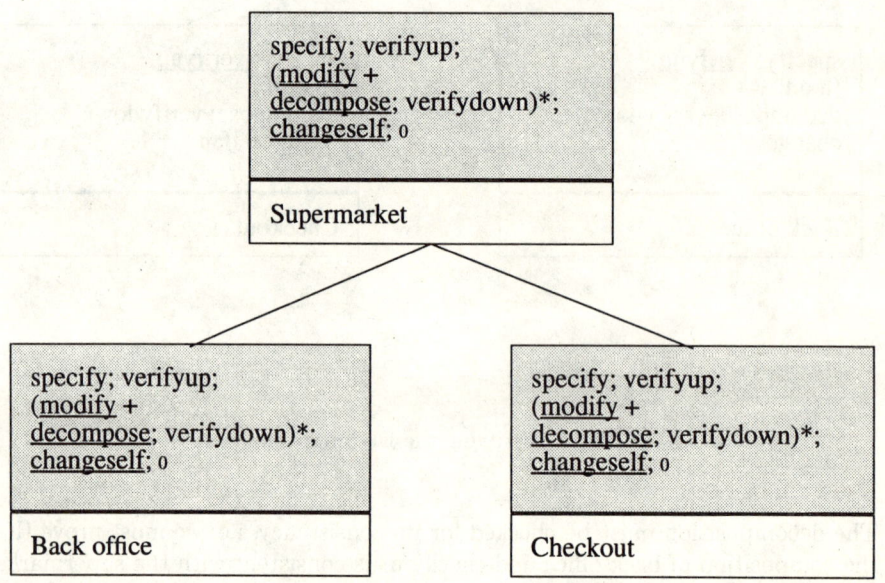

Fig. 6. After Modifying

After modifying the application, we again have the choice of doing either *modify*, *decompose* or *changeself*. See figure 6. Let's choose *changeself* to change the method part of the checkout node. For example, we may want to be able to execute the specification. So the new pattern for the method part may be

$$specify; verifyup; (modify \oplus execute \oplus decompose; verifydown)^*; changeself; \circ$$

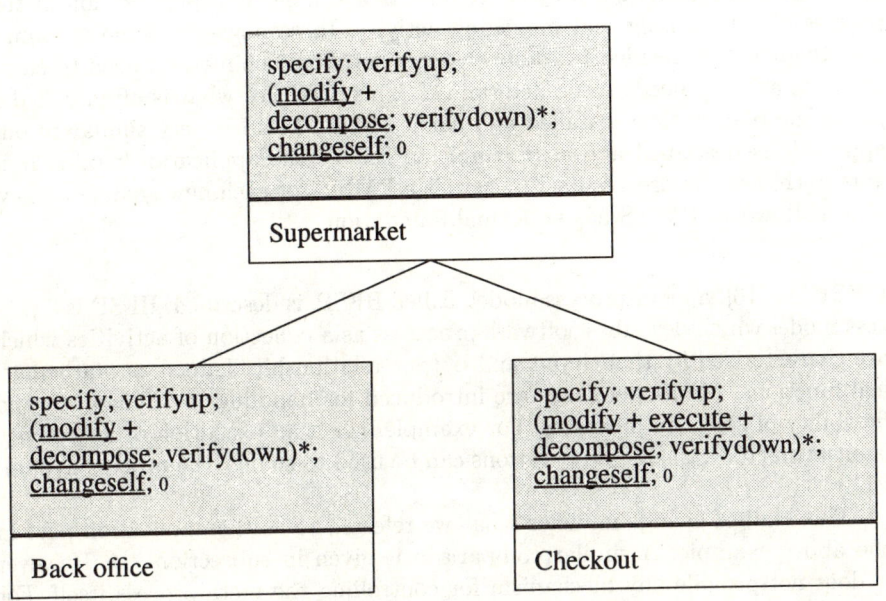

Fig. 7. New pattern for Checkout

In OBM, the meta-operation *changeself* for each node must ensure that the application being developed will remain consistent. The consistency check involves:

- re-specify the application part of the node;
- re-specify the application part of all the child nodes;
- prove that the composition of the new child specifications is consistent with the new specification of the node (according to the procedures described in [11]). The method part can only be changed if the consistency is maintained.

After a *changeself*, all the child nodes will inherit the new pattern. The currently enabled meta-operation will be identified. The child nodes will continue their development by following the new pattern.

8 Related Work

8.1 Meta-Processes

PMMS PMMS (Process Model for Management Support) [17][15] was developed under the IPSE 2.5 project. It reflects a simple conceptualization of the process of establishing and managing activity. In response to some stimulus, i.e. the goal, it is decided to establish some activity. The method used to carry out such activity needs to be determined. This produces what is often called a plan. The plan is then installed and carried out. PMMS is very similar to our approach as described above. It is able to change the application model. It is able to change the process itself. There is a PMMS for each new goal, i.e. a new object. However PMMS has no formal foundation.

HFSP In [16], a meta-process model, called HFSP, is described. HFSP is a process model which describes software processes as a collection of activities which are characterised by their input and output relationship defined as mathematical functions. Meta-operations are introduced for handling the dynamism and flexibility of software processes. For example, when an execution of an activity is unsatisfactory, the meta-operations can be used to change the process dynamically.

This kind of change modifies what we referred to as the application part in the above example. A similar comparison is given in subsection 7.4. However it does not provide any mechanism for controlling the meta-process itself. For example, if the meta-process is not suitable for a particular project, what can be done about it?

EPOS In [9], a meta-process is defined which is capable of changing different types of models, including the meta-model schema itself. The framework described in [9] does not provide formal support for validation or verification of process evolutions. Although in this paper, we have not discussed about how to validate or verify process changes, since the OBM language has formal semantics, it is possible to perform formal reasoning in our approach.

8.2 Specification of Interference

In [10], a method for specifying concurrent behaviour with possible interference is defined in terms of `rely` and `guarantee` conditions. The method is used to specify a system for shared variables. In OBM, communication and interaction between different components can only be achieved via operation calls.

The only possible interference comes from receiving results from an operation call to another component. Therefore, to address interference, it is only necessary to consider results received from the operation calls. This means that by slightly modifying the well known and widely accepted notation - *pre-* and *post*-conditions, we are able to embed the specification of interference.

8.3 Multiple Levels of Abstractions

The need for modelling processes in multiple levels of abstraction has been clearly recognised. Dowson pointed out in [7] that modelling formalism should accommodate a wide range of model granularity, and allow the refinement of initially large-grain models to address increasing details. The example above has shown that OBM has achieved this requirement.

Several existing approaches also provide means to represent a model in multiple levels of abstraction. Some of these are based on hybrid formalisms that use different notations for large-grain and small-grain aspects of process. An example of this type of approach is Process Weaver [8]. Another type of approach is incremental definitions. Examples of this type of approach include: tasks in Epos [5], activities in Slang [2], blackboard in Oikos [1]. However there is no support for formal reasoning between different levels of abstraction. Since Slang is based on Petri-nets, it should be possible to perform formal reasoning, however, it is not clear to the authors whether such a facility is provided by [2]. Work on consistency proof for refinement in Oikos is in development [1, 4].

In [12], a method based on data flow diagram (DFD) for modelling process interface is defined. Although the method has a formal basis for process decomposition, it provides very limited expressive power.

9 Conclusion

This paper has illustrated two main points. Firstly, it is possible to change an application process model dynamically in OBM. Secondly, it is possible to change the method through the OBM method itself. Our work on this issue is on-going, there is still a wide range of improvement and future work required. However we believe we have demonstrated that these two powerful issues can be addressed in a fairly simple framework.

References

1. V. Ambriola and Carlo Montangero. Oikos at the Age of Three. In *Proceedings EWSPT'92*, volume 635 of *Lecture Notes in Computer Science*. Springer Verlag, 1992.
2. S. Bandinelli, A. Fuggetta, and S. Grigolli. Process Modelling In-the-Large with SLANG. In *Proceedings of the Second International Conference on the Software Process*, Berlin, 1993.

3. H. Barringer, R. Kuiper, and A. Pnueli. Now you may compose temporal logic specifications. In *Proceedings of the 16th A.C.M. Symposium on Theory of Computing*, 1984.

4. X.J. Chen and C. Montangero. Compositional Refinement in Multiple Blackboard Systems. In *ESOP'92 European Symposium on Programming*, February.

5. R. Conradi et al. Design, Use and Implementation of SPELL, a Language for Software Process Modeling and Evolution. In *Proceedings EWSPT'92*, volume 635 of *Lecture Notes in Computer Science*. Springer Verlag, 1992.

6. R. Conradi, C. Fernström and A. Fuggetta. A Conceptual Framework for Evolving Software Processes. in *Software Process Modelling and Technology*, A. Finkelstein, J. Kramer and B. Nuseibeh (Eds.), Research Studies Press, Wiley, 1994.

7. Mark Dowson. Software Process Themes and Issues. In *Proceedings of the Second International Conference on the Software Process*. IEEE Computer Society Press, 1993.

8. C. Fernström. PROCESS WEAVER: Adding Process Support to UNIX. In *Proceedings of the Second International Conference on the Software Process*, Berlin, 1993.

9. M.L.Jaccheri and R.Conradi. Techniques for Process Model Evolution in EPOS. in IEEE TSE.

10. C.B. Jones, Tentative Steps Toward a Development Method for Interfering Programs, *ACM TOPLAS* 5(4), 1983.

11. J.A.Keane, J.Sa and B.C.Warboys, Applying a Concurrent Formal Framework to Process Modelling. In *Proceedings of FME'94*, Barcelona, October, 1994. Pages 291-305, LNCS 873.

12. C. Kung. Process Interface Modelling and Consistency Checking. *Journal of System and Software*, 15:185-191, 1991.

13. L. Lamport. What Good is Temporal Logic? In R.E.A. Mason, editor, *Information Processing 83*, pages 657-668, IFIP, 1983.

14. J. Sa and B.C. Warboys. Specifying Concurrent Object-based Systems using Combined Specification Notations. Technical Report UMCS-91-9-2, Department of Computer Science, University of Manchester, July 1991.

15. R. Snowdon. An Example of Process Change. In *Proceedings EWSPT'92*, volume 635 of *Lecture Notes in Computer Science*, pages 179–195. Springer Verlag, 1992.

16. M.Suzuki and T.Katayama. Meta-Operations in the Process Model HFSP for the Dynamics and Flexibility of Software Processes. in Proceedings of the First International Conference on software Processes, 1991, pages 202–217.

17. B. Warboys. The IPSE2.5 Project: Process Modelling as the Basis for a Support Environment. In *Proceedings of the First International Conference on System Development Environment and Factories*, Berlin, May 1989

18. P. Wegner, Dimensions of Object-Based Language Design. in *OOPSLA'87 Proceedings*, October, 1987.

Transients Change Processes in Process Centered Environments

Ali B. Kaba and Jean-Claude Derniame

CRIN-Bâtiment Loria
BP 239 - 54506 Vandoeuvre-Les-Nancy
e-mail:kaba@loria.fr, derniame@loria.fr

1 Introduction

Recent works on software process evolution [Belkhatir 92, Bandinelli 93, Jaccheri 93, Kaiser 93, Kaba 94] focus on the evolution of process model. But evolution of the process itself needs also specific support. This kind of evolution involves not only to identify the new requirements, and to built the new process but also to built a detailed meta-process describing the migration from the old process to the new one.

This paper proposes a way to migrate an existing process to a target process. It emphasises on dynamic and evolutionary changes which allow us to perform a partial or a full transformation of a process behaviour. Therefore, our attention concentrates on two main aspects.

First, we assume that changes are performed by a meta-process. To be independent of each process instance, general rules have to be abstracted from usual changes to built the meta-process.

Next, the transient process should be defined in a way such that it's execution leave the environment in a consistent state, without causing disturbance to the unaffected enacting processes.

The approach outlined in this paper uses concepts developed in[Derniame 94].

2 Objectives and Goals

It will be common for a project to change its process definition during the enactment of that process: it is quite possible to begin enactment of some parts of a process when the others parts still have to be defined, and equally, the need to define an existing definition may arise at any time. To accommodate developing and enacting processes with these kinds of changes, the process support technology must be sufficiently flexible to permit arbitrary incremental change. Moreover, the process support technology should be capable to perform such evolutions dynamically by disturbing or interrupting just a minimal set of others enacting processes.

The majors goals for transients processes are:

- to provide mechanisms for migrating a process towards the target process.
- to guide the preparation of the target environment,
- to proof :
  - that the target process is attainable from the existing process and,
  - that the evolution should leave the system in a consistent state,
- to check behavioral knowledge at the transition time.

We discuss below how to reach these goals.

3 Transients Changes Management

It has been argued[Conradi-Fuggetta 94] that there are numerous advantages of creating an explicit model of the transient process. In order to provide a sound basis for transients change management we describe the concept of activity. Then, we present the form of interaction between Transients change models and the activities. This leads to the transient derivation mechanism.

3.1 Activities vs transients processes interaction

An activity is a production step of a process. Its definition is composed of an activity instances, its name, and its relationships with product, direction, tool, role, and sub-activity instance. A defined activity can be carried out when products are under development on the process support environment.

When changing an activity, the process support should direct the activity toward an appropriate state for the transient change performance. Activities states and the transitions to reach these states are described below(see figure 1). The invariants that transition operations must satisfy, in order to change activities states are described in [PCIS 94].

Activity states are defined such as *non-existent, described, established, active, suspended, finished, closed*. When *established* an activity is defined, and assigned to a role, it waits only for an agent to start it. Being in this state, an activity is consistent and frozen[1], and it is amenable for enactment. The transient change process mechanism interacts with the enacting process by performing a partially ordered set of actions. The transient change process mechanism is described in section 3.2.

---

[1] An activity state is frozen if he don't perform currently a work, so its related entities are stable.

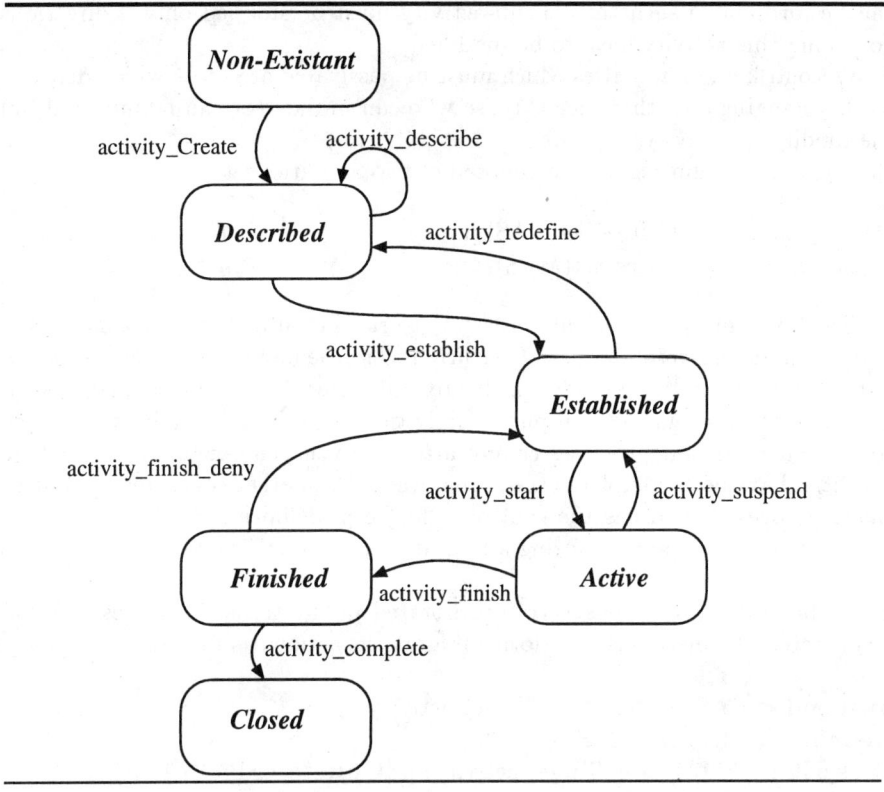

Fig. 1. States of an activity

3.2 Transient Change Process derivation mechanism

The transient process derivation mechanism is defined as follows:

- It establishes an *inaction region* around the target activity.
  The activities members of the inaction region are those which can initiate (or can be initiated by) the target activity.
- It performs an ordered set of activities on the inactive region in order to modify the concerned activities behaviour, all in checking the operations invariants in conformance to the target activity description.

When establishing an inaction region, we distinguish two kinds of activities which must be inactive.
The first set contains activities definition which should be modified during the change. For instance, when applying a full change policy, all the sub-activities of the changing activity will be included in this set. This set can be considerably reduced if the change is partial. Partial changes depend of the local policy enhanced for a given product. For example in LCPS-PCIS[Derniame 94], a policy

can be formulated such that, a sub-activity must be stopped only if directions governing this activity have to be modified.

The second kind of activities which must be passivated are those which depends to the changing one, that means those who can initiate (or can be initiated by) the modifying activity.

Briefly, the inaction region is composed of two activities sets:

- Dependants Activities Set(DAS) and,
- Modifiable Activities Set(MAS).

To allow multiple transients processes to run in parallel, we use concurrents *privates inactives activities sets*. Concurrents transients processes has also *semi-public inactives activities sets* which are supersets of concurrents privates inactives activities sets. A semi-public activities set includes activities on which concurrent migrating processes cannot attempt to suspend or to remove if these activities have been completed (resp. removed)[2] by another concurrent transient change process which has the control right (resp. definition right) on them. A semi-public set is updated during a migrating process description.

The sets as described above are incorporated in the transient process mechanism state. The mechanism performs migrations actions as follows:

Suspend $< DAS + MAS > +$ Target activity
Redefine $< MAS > +$ Target activity
(**Describe** $< (MAS > +$ Target activity) $|| [$**Update** $<SPA^3>])^*$
Establish $< MAS > +$ Target activity
Start $< MAS - SAS> +$ Target activity
Resume $< DAS - SAS >$

4 Conclusion

The key concept in Transient Change Process management is an attempt to abstract an enacting process state into a migration process state which forms the basis for the transient derivation mechanism. The study of multiple Transients processes is an important issue. Also, the approach as presented is independent of the migrating process granularity. We defined simple composition rules which provides coarser grains for transient process management. For a finer grain, decomposition can be used to expose the internal structure of interconnected processes to make them accessible to change.

We are currently exploring the use of LCPS-PCIS[PCIS 94] environment to implement our approach.

---

[2] This set will be called latter: Stopped Activities Set(**SAS**)
[3] Semi-Public Activities Set(**SPA**)

References

[Bandinelli 93] **Bandinelli S., Fuggetta A., Ghezzi C.** *Process Model Evolution in the SPADE Environment* IEEE Transactions on Software Engineering. Special Issue on Process Evolution December 1993.

[Belkhatir 92] **Belkhatir N. & al** *Supporting software maintenance evolution processes in the Adele system* Proc. of the 30th Annual ACM Southeast Conf. pp 165–172 Apr. 92.

[Conradi-Fuggetta 94] **Conradi R., Fernstrom C., Fuggetta A.** *A Conceptual Framework for Evolving Software Processes* ACM Sigsoft Software Engineering Notes Vol 13 No 4 Oct 93.

[Jaccheri 93] **Jaccheri M.L. and Conradi R.** *Techniques for Process Model Evolution in EPOS.* IEEE Trans. on Software Engineering , December 1993.

[Kaba 94] **Kaba A. B., Tankoano J., Derniame J-C** *Une approche incrémentale d'évolution des modèles de procédés de développement de logiciels* CARI'94 proceedings.

[Kaiser 93] **Kaiser G., Ben-Shaul I. Z.** *Process Evolution in the MARVEL Environment* Proceed. of Eighth International Software Process Workshop (ISPW8),Dagstuhl, Germany, March 1993.

[Derniame 94] **Derniame J. C. & All** *Life Cycle Process Support in PCIS* To appear in PCTE'94 proceedings.

[Tankoano 94] **Tankoano J., Derniame J-C, Kaba A. B.** *Software Process Design Based on Products and the Object Oriented Paradigm* in Warboys B. C. Soft. Process Techn. LNCS No 772. Springer-Verlag Feb 94.

[PCIS 94] **PCIS-WG** *LCPS-PCIS (DA4/DA5)*

Author Index

Springer-Verlag
and the Environment

We at Springer-Verlag firmly believe that an international science publisher has a special obligation to the environment, and our corporate policies consistently reflect this conviction.

We also expect our business partners – paper mills, printers, packaging manufacturers, etc. – to commit themselves to using environmentally friendly materials and production processes.

The paper in this book is made from low- or no-chlorine pulp and is acid free, in conformance with international standards for paper permanency.

Lecture Notes in Computer Science

For information about Vols. 1–835
please contact your bookseller or Springer-Verlag

Vol. 871: J. P. Lee, G. G. Grinstein (Eds.), Database Issues for Data Visualization. Proceedings, 1993. XIV, 229 pages. 1994.

Vol. 872: S Arikawa, K. P. Jantke (Eds.), Algorithmic Learning Theory. Proceedings, 1994. XIV, 575 pages. 1994.

Vol. 873: M. Naftalin, T. Denvir, M. Bertran (Eds.), FME '94: Industrial Benefit of Formal Methods. Proceedings, 1994. XI, 723 pages. 1994.

Vol. 874: A. Borning (Ed.), Principles and Practice of Constraint Programming. Proceedings, 1994. IX, 361 pages. 1994.

Vol. 875: D. Gollmann (Ed.), Computer Security – ESORICS 94. Proceedings, 1994. XI, 469 pages. 1994.

Vol. 876: B. Blumenthal, J. Gornostaev, C. Unger (Eds.), Human-Computer Interaction. Proceedings, 1994. IX, 239 pages. 1994.

Vol. 877: L. M. Adleman, M.-D. Huang (Eds.), Algorithmic Number Theory. Proceedings, 1994. IX, 323 pages. 1994.

Vol. 878: T. Ishida; Parallel, Distributed and Multiagent Production Systems. XVII, 166 pages. 1994. (Subseries LNAI).

Vol. 879: J. Dongarra, J. Waśniewski (Eds.), Parallel Scientific Computing. Proceedings, 1994. XI, 566 pages. 1994.

Vol. 880: P. S. Thiagarajan (Ed.), Foundations of Software Technology and Theoretical Computer Science. Proceedings, 1994. XI, 451 pages. 1994.

Vol. 881: P. Loucopoulos (Ed.), Entity-Relationship Approach – ER '94. Proceedings, 1994. XIII, 579 pages. 1994.

Vol. 882: D. Hutchison, A. Danthine, H. Leopold, G. Coulson (Eds.), Multimedia Transport and Teleservices. Proceedings, 1994. XI, 380 pages. 1994.

Vol. 883: L. Fribourg, F. Turini (Eds.), Logic Program Synthesis and Transformation – Meta-Programming in Logic. Proceedings, 1994. IX, 451 pages. 1994.

Vol. 884: J. Nievergelt, T. Roos, H.-J. Schek, P. Widmayer (Eds.), IGIS '94: Geographic Information Systems. Proceedings, 1994. VIII, 292 pages. 19944.

Vol. 885: R. C. Veltkamp, Closed Objects Boundaries from Scattered Points. VIII, 144 pages. 1994.

Vol. 886: M. M. Veloso, Planning and Learning by Analogical Reasoning. XIII, 181 pages. 1994. (Subseries LNAI).

Vol. 887: M. Toussaint (Ed.), Ada in Europe. Proceedings, 1994. XII, 521 pages. 1994.

Vol. 888: S. A. Andersson (Ed.), Analysis of Dynamical and Cognitive Systems. Proceedings, 1993. VII, 260 pages. 1995.

Vol. 889: H. P. Lubich, Towards a CSCW Framework for Scientific Cooperation in Europe. X, 268 pages. 1995.

Vol. 890: M. J. Wooldridge, N. R. Jennings (Eds.), Intelligent Agents. Proceedings, 1994. VIII, 407 pages. 1995. (Subseries LNAI).

Vol. 891: C. Lewerentz, T. Lindner (Eds.), Formal Development of Reactive Systems. XI, 394 pages. 1995.

Vol. 892: K. Pingali, U. Banerjee, D. Gelernter, A. Nicolau, D. Padua (Eds.), Languages and Compilers for Parallel Computing. Proceedings, 1994. XI, 496 pages. 1995.

Vol. 893: G. Gottlob, M. Y. Vardi (Eds.), Database Theory – ICDT '95. Proceedings, 1995. XI, 454 pages. 1995.

Vol. 894: R. Tamassia, I. G. Tollis (Eds.), Graph Drawing. Proceedings, 1994. X, 471 pages. 1995.

Vol. 895: R. L. Ibrahim (Ed.), Software Engineering Education. Proceedings, 1995. XII, 449 pages. 1995.

Vol. 896: R. N. Taylor, J. Coutaz (Eds.), Software Engineering and Human-Computer Interaction. Proceedings, 1994. X, 281 pages. 1995.

Vol. 897: M. Fisher, R. Owens (Eds.), Executable Modal and Temporal Logics. Proceedings, 1993. VII, 180 pages. 1995. (Subseries LNAI).

Vol. 898: P. Steffens (Ed.), Machine Translation and the Lexicon. Proceedings, 1993. X, 251 pages. 1995. (Subseries LNAI).

Vol. 899: W. Banzhaf, F. H. Eeckman (Eds.), Evolution and Biocomputation. VII, 277 pages. 1995.

Vol. 900: E. W. Mayr, C. Puech (Eds.), STACS 95. Proceedings, 1995. XIII, 654 pages. 1995.

Vol. 901: R. Kumar, T. Kropf (Eds.), Theorem Provers in Circuit Design. Proceedings, 1994. VIII, 303 pages. 1995.

Vol. 902: M. Dezani-Ciancaglini, G. Plotkin (Eds.), Typed Lambda Calculi and Applications. Proceedings, 1995. VIII, 443 pages. 1995.

Vol. 903: E. W. Mayr, G. Schmidt, G. Tinhofer (Eds.), Graph-Theoretic Concepts in Computer Science. Proceedings, 1994. IX, 414 pages. 1995.

Vol. 904: P. Vitányi (Ed.), Computational Learning Theory. EuroCOLT '95. Proceedings, 1995. XVII, 415 pages. 1995. (Subseries LNAI).

Vol. 905: N. Ayache (Ed.), Computer Vision, Virtual Reality and Robotics in Medicine. Proceedings, 1995. XIV, 567 pages. 1995.

Vol. 906: E. Astesiano, G. Reggio, A. Tarlecki (Eds.), Recent Trends in Data Type Specification. Proceedings, 1995. VIII, 523 pages. 1995.

Vol. 907: T. Ito, A. Yonezawa (Eds.), Theory and Practice of Parallel Programming. Proceedings, 1995. VIII, 485 pages. 1995.

Vol. 908: J. R. Rao Extensions of the UNITY Methodology: Compositionality, Fairness and Probability in Parallelism. XI, 178 pages. 1995.

Vol. 910: A. Podelski (Ed.), Constraint Programming: Basics and Trends. Proceedings, 1995. XI, 315 pages. 1995.

Vol. 911: R. Baeza-Yates, E. Goles, P. V. Poblete (Eds.), LATIN '95: Theoretical Informatics. Proceedings, 1995. IX, 525 pages. 1995.

Vol. 913: W. Schäfer (Ed.), Software Process Technology. Proceedings, 1995. IX, 261 pages. 1995.

Vol. 914: J. Hsiang (Ed.), Rewriting Techniques and Applications. Proceedings, 1995. XII, 473 pages. 1995.